Industrial Applications of Surfactants III

Industrial Applications of Surfactants III

Edited by
D. R. Karsa
Harcros Chemicals (UK) Limited, Manchester

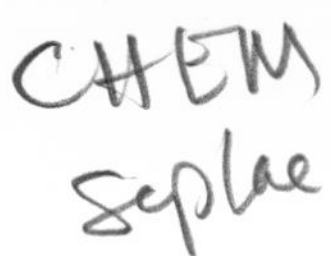

The Proceedings of a Symposium organized by the North West Region of the Industrial Division of the Royal Society of Chemistry. University of Salford, 16–18th September 1991.

Special Publications No. 107

ISBN 0-85186-227-6

A catalogue record of this book is available from the British Library

© The Royal Society of Chemistry 1992

All Rights Reserved
No part of this book may be reproduced or transmitted in any form or by any means—graphic, electronic, including photocopying, recording, taping, or information storage and retrieval systems—without written permission from The Royal Society of Chemistry

Published by The Royal Society of Chemistry,
Thomas Graham House, Science Park, Cambridge CB4 4WF

Printed in England by Redwood Press Ltd, Melksham, Wiltshire

TP994
I53
1992
CHEM

Preface

This book is a summary of the third Royal Society of Chemistry international symposium on the "Industrial Applications of Surfactants", the previous meetings being held in 1986 and 1989 respectively. Surfactants belong to many chemical classes and are used across the whole spectrum of industry. Hence, any two-day meeting can only provide a snapshot of selected products and application areas. As with the previous two meetings, completely new products and applications are described, the contents of each meeting being complementary to the others.

Where possible, speakers were asked to emphasise the more practical aspects of their products. Once again one or two of the newer more innovative areas, such as polymerisable surfactants and comb surfactants have been highlighted, together with further descriptions of surfactants under their various ionic classifications. Here, likewise, one or two less well-known and newer surfactant types are described, including sarcosinates, sulphobetaines, and alkyloxypropylamines. Finally a limited number of application areas are discussed, including oil field chemicals, emulsifiers for the food industry and new hydrotropes of built liquids.

It is hoped that this volume will prove to be a useful addition to the surfactant bibliography and of interest to those involved in the production, formulation and application of surfactant-based products.

D R KARSA

Contents

Nonionic Surfactants

Cationic and Amphoteric Surfactants

Applications

Introduction

The Market for Industrial Surfactants in the 1990s — An Overview

Alan H. Turner and Joel H. Houston

COLIN A. HOUSTON & ASSOCIATES INC., P.O. BOX 416, MAMARONECK, NY 10543 USA

Introduction

My brief is to give an overview of the market for industrial surfactants in the 1990s. This is an enormous subject, and all I can really hope to do is to highlight what seem to me to be some of the main features in West Europe and in North America.

We are all very familiar with the use of surfactants in household detergents. Our television screens remind us of this every evening. The industrial applications are not so obvious; they are much more diverse, but they represent an enormous tonnage.

Perhaps I can bring this home to you in the following way. Many of you will have come here by car. Where is the link between cars and surfactants, apart from the obvious one of the car wash? Think of the manufacturing process: surfactants used in metal cleaning and metal working, in the paint, in the wax polish, in the manufacture and dyeing of the seat and carpet fabrics. The rubber in the tires will probably have been made in an emulsion polymerization process, which depends on surfactants. Let us not forget the detergent additives in the fuel, the lubricants, the grease...

If it is a luxury car, there will be some leather in it - more surfactants used here. Is there a foam fire extinguisher?

But let's get out of the car. The asphalt of the car park makes us think of bitumen emulsions, and the concrete of super-plasticizers and foamed concrete. The weeds are kept under control with agrochemicals, perhaps supplied as an emulsion concentrate?

If like me you had a sandwich and a glass of beer for your

lunch, you may have thought of the food emulsifiers used in bread manufacture, the surfactants used for glass and bottle washing, and then the bill - the paper industry uses surfactants too. I could go on, but I am sure you get my point. Is there any major manufacturing industry which does NOT use any surfactants somewhere, disguised as a process aid? Can anyone think of one?

Something else that may strike you: most of these applications have been with us for many years. Emulsion polymerization was developed during World War II, and the use of synthetic surfactants in the textile industry goes back to the 1930s. Metal cleaning and working, leather and paper manufacture, food, paint - these are long-established industries with long-established processes. In most cases (though not quite all), the types of surfactants they are using today are the same ones they have been using for many years. This is often because of the sheer amount of work (and, of course, associated cost) of evaluating alternative surfactants. Where are the new applications? What is the incentive for a surfactant manufacturer to look towards the industrial applications sector? What are the prospects for novel surfactants?

This is the third, well-attended, Salford conference on industrial surfactants, so clearly there is a lot of interest and activity in this subject. I am sure that we shall hear interesting things about novel surfactants and novel applications in the course of this symposium. Let us remember that today's novel surfactant application may - in favorable circumstances - grow into tomorrow's commodity business.

Industrial Surfactants 1990 - West Europe and U.S.

But let us return to 1991. I shall give some figures to show the size of the industrial surfactant market in W. Europe and the most important types of surfactants used.[1] I shall also show for comparison some corresponding data for the U.S., taken from a study which our company issued in July this year.[2]

We have to be careful here with our definitions. What do we include as an industrial use? Does it mean all non-domestic applications, including institutional cleaning? What do we include as a surfactant? Do we include soap or not? Some writers do not consider lignosulfonates to be true surfactants. What about detergent additives for lubricating oil, or frothing agents for ore flotation? These are matters of judgment, but they need to be borne in mind when comparing figures from different sources. Figure 1 shows the consumption of surfactants (including soap) in industrial applications in West Europe and the U.S. in 1990, and the main types of surfactants involved.

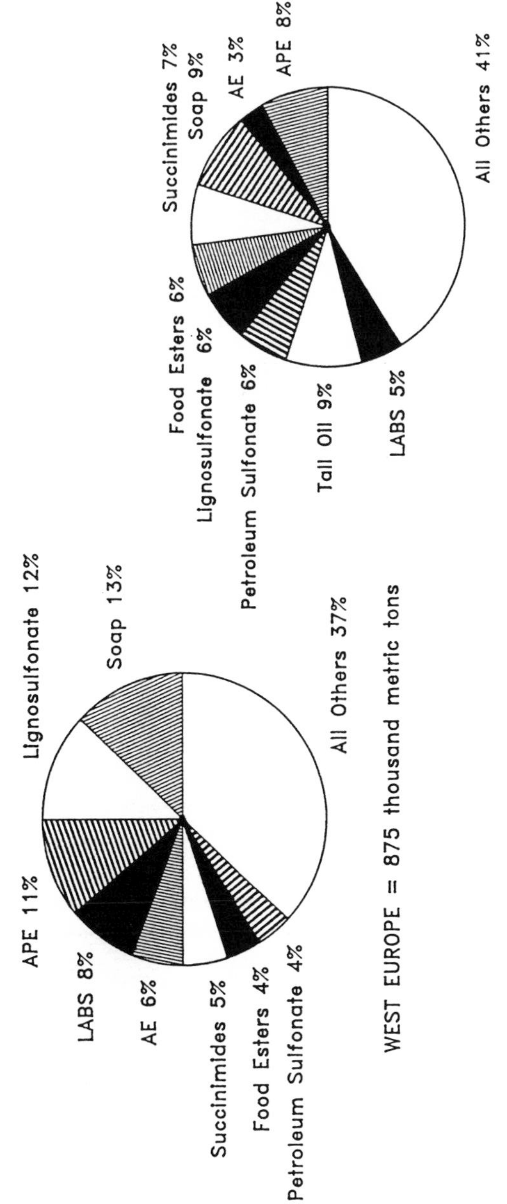

Figure 1 SURFACTANTS IN INDUSTRIAL APPLICATIONS, 1990

When we look at the types of surfactants being used in West Europe, we have a total surfactant consumption of 875 thousand tons, with soaps, lignosulfonates and APE as the most important materials, followed by LABS, AE, succinimides (entirely for petroleum additives), glycerol and glycol esters, and petroleum sulfonates.

In the U.S., the total is much larger at 1470 thousand tons. Soaps, lignosulfonates and APE are all important but we also have tall oil derivatives which make up nearly 10% of the total, but which are negligible in Europe. The "Others" category includes just about every type of surfactant commercially available.

Figure 2 shows the relative importance of industrial applications in the total consumption of surfactants in West Europe and the U.S. These figures relate to 1990 and do NOT include soaps. In volume terms, the total consumption of surfactants in household applications and in personal care products is approximately the same in West Europe and in the U.S. However, industrial outlets in the U.S. take a significantly bigger share of the pie (and it is a large pie) than they do in West Europe. For other parts of the world, household outlets would take an even larger share, and industrial outlets a smaller share, than in West Europe.[3]

As an approximation, about half the total surfactant consumption in the U.S., or more than 1 million tons, goes into industrial outlets. For West Europe, industrial outlets account for around one-third of the total surfactant consumption (excluding soap).

It is not too difficult to understand why the industrial use of surfactants in the U.S. should be so high. After I&I cleaning, the largest single use sector is in the petroleum industry. Figure 3 shows the various ways in which surfactants are used in the petroleum industry. In terms of volume, lube oil additives are dominant; we find here large volumes of specialty surfactants which have little or no use in other applications, e.g. the polybutene succinimides, alkyl phenates and salicylates. With the highly developed state of the petroleum industry and dependence on the automobile, it is not surprising that nearly a quarter of a million tons of surfactants are consumed each year in the U.S. as lubricating oil additives, making this the largest single industrial outlet for surfactants. Significant quantities of surface-active materials are also used in petroleum exploration and production activities[4].

Petroleum additives are not quite so important in West Europe, where they account for about 11% of the total industrial surfactants consumption, as against 17% in the U.S. (Figure 4). On the other hand, the plastics and elastomers industry takes a

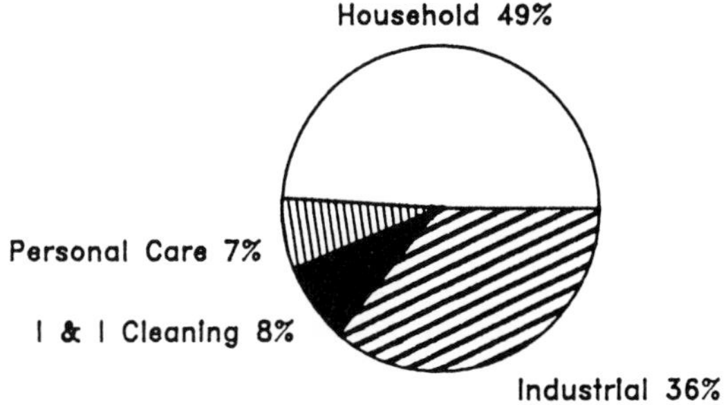

WEST EUROPE = 1800 thous. metric tons

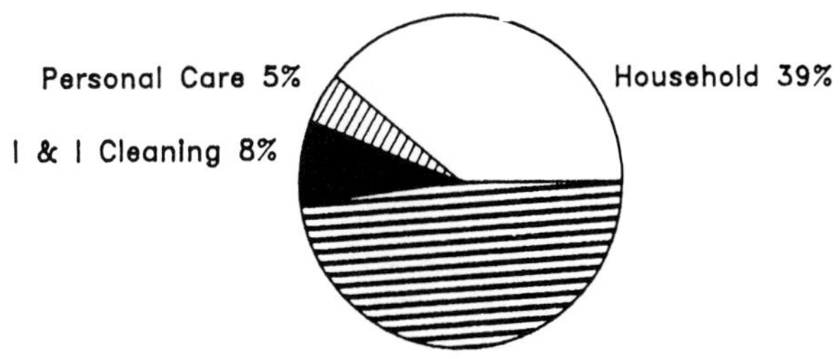

UNITED STATES = 2210 thous. metric tons

Figure 2 SURFACTANT* OUTLETS IN WEST EUROPE & U.S.

* Does not include soap.

- Lube Oil Additives
- Fuel Additives
- Grease Additives
- Emulsifiers/Demulsifiers
- Biocides
- Drilling Mud Additives
- Corrosion Inhibitors
- Enhanced Oil Recovery
- Others

Figure 3 USES OF SURFACTANTS IN THE PETROLEUM INDUSTRY

rather larger share of the total industrial surfactants consumption in West Europe (14%), than it does in the U.S. (8%). The main application here is emulsion polymerization, although use as antistatic additives in a finished polymer is of some importance. Various types of soaps are important here.

The I&I cleaning markets are important both in the U.S. (15%) and in Europe (18%).

The second largest industrial outlet for surfactants in the U.S. is ore flotation, which is understandably more important in the U.S. (11%) than it is in Europe. The function of surfactants in ore flotation is to separate unwanted material from desirable ore components. The desirable minerals are captured by surfactant froth bubbles and so rise to the top of the bath. Chemicals which assist by increasing the volume of foam ("frothers") include methylisobutyl carbinol, linear alcohols and polypropylene glycol, none of which is normally regarded as a surfactant. Agents which promote incorporation of the desired mineral in the froth bubbles are called "collectors," and include tall oil fatty acids, amines and xanthates.

In the U.S., the food industry accounts for around 7% of the total industrial consumption. Surfactant additives are used in foods both as emulsifiers (to modify starch, protein or fatty products) and to improve texture and reduce "staling," as well as improving finished product quality. Needless to say, only a very restricted range of surfactants, particularly glycerides, are approved for use in food. Perhaps in Europe we are not so fond of having our food improved: surfactant use in the food industry is relatively modest at around 5% (<u>Figure 4</u>).

One sector where Europe is ahead of the U.S. is the textile industry (12% of industrial surfactants in W. Europe, only 6% in the U.S.)

Textiles is really a group of industries, including natural fibers such as cotton and wool as well as the production and processing of synthetic fibers (rayon, polyester, polyolefin, etc.) Virtually every type of surfactant finds some use in the textile industry. Uses include the scouring of raw cotton and wool, aids for applying and removing other process chemicals, and in synthetics as fiber lubricants, scouring, weaving, dyeing and printing aids, as well as assistants for the application of antistatic, permanent press and other finishes.

Another industry where surfactant consumption is more important in Europe than in the U.S. is cement and concrete, where surfactants are used as air-entraining agents and as components of

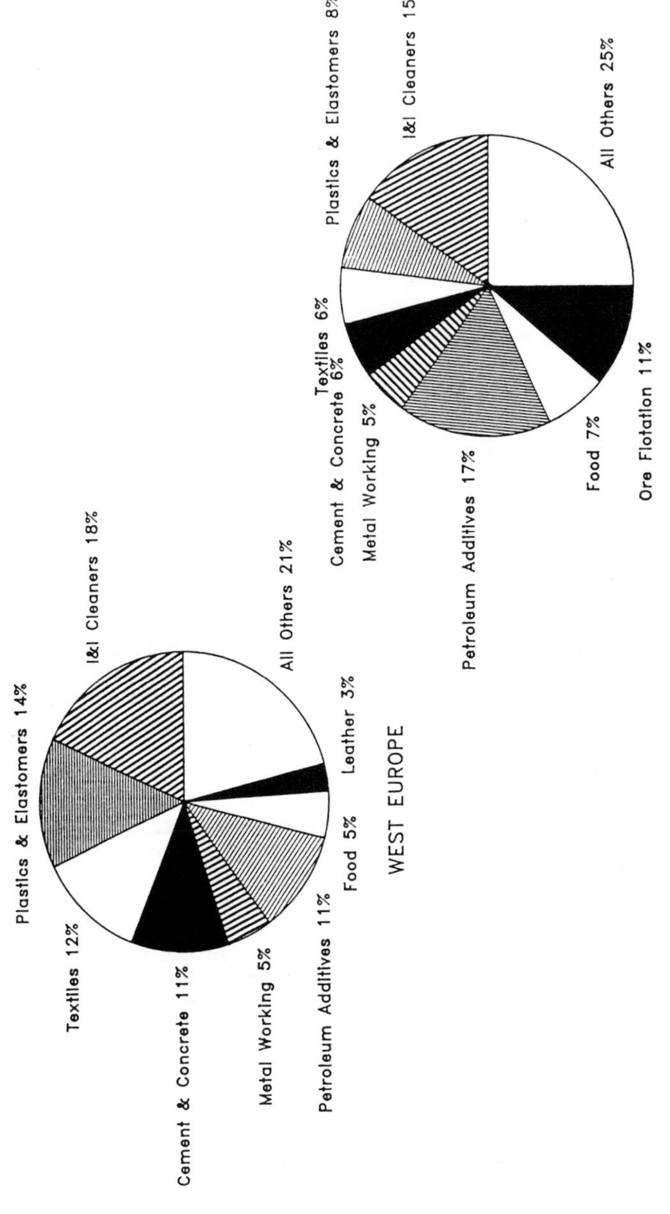

Figure 4 MAJOR INDUSTRIAL OUTLETS FOR SURFACTANTS, 1990

admixtures (superplasticizers, water reducing agents, etc.) (11% in W. Europe, only 6% in U.S.)

A lot of lignosulfonates are used, but synthetic sulfonates based on naphthalene/formaldehyde or melamine/formaldehyde condensates are becoming increasingly popular.

The metalworking industry uses a lot of surfactants, not as foaming agents, but in cutting fluids, rolling oils, drawing compounds and forging lubricants. The surfactants act as emulsifiers, lubricants and corrosion inhibitors, (5% in both W. Europe and the U.S.).

Time does not allow me to comment on all the industrial uses, and I would remind you that there are many interesting applications hidden in the "Others" category of Figure 4.

Why are surfactants used in these various applications? In general, the role of the surfactant in these various processes derives from its ability to undergo adsorption at an interface. This can result in the lowering of a surface or interfacial tension, and/or a stabilization of one or more interfaces (as for example in a foam or an emulsion). Solubilization can also be important. It is perhaps easy to understand in a qualitative way how a surfactant can generate or stabilize a foam, and lubricate or wet a surface, but the question of why one type of surfactant functions better than another type in a specific application is much more complex. Indeed, it is commonly held that the selection of a surfactant for an application such as emulsion polymerization is more of an art than a science.

For a surfactant producer, the household sector and the industrial sector call for very different approaches. In the household sector, you have to deal in the main with a handful of multinational customers who tell their suppliers in broad terms what they want, how much (or usually how little!) they are prepared to pay, but not how they intend to use the product. In the industrial sector, the customer is often not an expert in surfactants, and is looking for an "effect." The cost of the surfactant is not unimportant, but often it is a minor item in the total process cost. The customer does expect the surfactant producer to know all about his product, and also to be expert in the customer's own application and be able to assist in sorting out any problems. The supplier has to be service-oriented. While, as noted a moment ago, the cost of the surfactant may be relatively unimportant, if the overall process is upset, the operating cost can be enormous.

No surfactant supplier can be expert in all industrial application areas, so we have seen the development of companies which specialize in supplying a range of auxiliaries (not only surfactants but also other chemicals) to a specific industry. Figures 5-9 show some examples of companies which engage in this type of activity in the plastics and elastomers, petroleum, textile, cement and concrete, and food industries. These are a representative selection and not intended to be comprehensive; we apologize to any companies which have been left out.

Some of the largest industrial users of surfactants are themselves surfactant producers, who can develop processes and applications based on their own surfactant range. Figure 10 gives some examples and again, this list does not pretend to be complete.

Although these producers do not necessarily all make a complete range of surfactants, a company such as ICI would have little need to look beyond their own product range for their surfactant requirements.

Future Prospects for Industrial Surfactants

So much for the industrial surfactant market as it stands today. Where do we go from here? Many of the applications I have mentioned have been around now for a good many years. How is industry going to respond to the challenge to come up with some new applications and some new surfactants? We can consider the future outlook under three headings: the overall surfactant requirements of individual industries, the replacement of current surfactants by others, and novel applications.

a) Overall requirements by individual industries

We have seen in recent times a period of low economic growth which has had a major impact on industries such as textiles, metal, construction etc. This has of necessity been reflected in the consumption of surfactants by these industries. There is some hope that things are beginning to improve slowly, but the overall prospects are far from exciting. The textile industry is vulnerable to imports, while the leather industry continues a slow decline. Industries such as cement and concrete, metalworking, or asphalt tend to run parallel with the national economy of the country.

b) Replacement of current surfactants by others?

One of the driving forces for change is of course the regulatory issue, and, in particular, concern about effluents. When an aqueous effluent has to be disposed of via municipal waste water treatment plants, there is natural concern about biodegradability, foaming and aquatic toxicity. Not all industrial uses of surfactants generate an aqueous effluent, but those which do (e.g. textiles, paper, leather) are under increasing pressure to reduce and clean up their effluents.

- Akzo Chemie BV
- Auschem SpA
- Harcros Chemicals UK Ltd.
- Rhone Poulenc
- Schill & Seilacher GmbH
- Witco SA

Figure 5 MAJOR CHEMICAL SUPPLIERS FOR PLASTICS & ELASTOMERS

- Ethyl Petroleum Additives Ltd
- Exxon
- Lubrizol
- Petrolite Ltd

Figure 6 MAJOR CHEMICAL SUPPLIERS FOR PETROLEUM ADDITIVES

- Allied Colloids PLC
- Dr. Th. Boehme KG
- Crosfield Chemicals
- Yorkshire Chemicals Ltd
- Zschimmer & Schwarz GmbH

Figure 7 MAJOR CHEMICAL SUPPLIERS FOR TEXTILES

- W.R. Grace Ltd
- Henkel AG
- Hoechst AG
- Lignotech
- Westvaco

Figure 8 MAJOR CHEMICAL SUPPLIERS FOR CEMENT & CONCRETE

- Croda
- Grindsted Products A/S
- Th. Goldschmidt AG
- Quest Int'l

Figure 9 MAJOR CHEMICAL SUPPLIERS FOR FOOD

- Akzo (Asphalt, Paint)
- Dow (Agrochemicals, Polymers)
- Hoechst (Polymers, Textiles)
- Hüls (Polymers)
- ICI (Paint, Polymers, Agrochemicals)
- Rhône Poulenc (Agrochemicals, Polymers, Textiles)
- Shell (Polymers, Agrochemicals, Petroleum)

Figure 10 MAJOR COMPANIES WHICH BOTH PRODUCE AND USE INDUSTRIAL SURFACTANTS

Figure 11 shows some industrial surfactants which are being replaced, or which may have to be replaced in the future, for environmental reasons. Thus APE, which is one of the most widely used surfactants in the industrial sector, is progressively being replaced by more environmentally acceptable surfactants. In West Germany (as it then was), after discussions between the detergent industry and the Government, use of APE was reduced on a voluntary basis starting in 1984 until today it is only used on a limited basis in those industrial applications which give rise to no aqueous effluent. In Switzerland, a law was passed in 1986 to prohibit the use of APE specifically in the textile industry. It appears likely that the use of APE by the textile industry in Italy may soon be regulated or even banned. An important factor in these developments was the availability of alternative surfactants with better environmental properties and (more or less) equivalent performance to APE. Without going into the pros and cons of the case against APE, I think we must expect the use of this surfactant to continue to decline in West Europe, at least in those applications which generate an aqueous effluent.

The biodegradability (under real world conditions) of other surfactants may also be questioned in the present climate of environmental sensitivity. There is, I think, no question that EO/PO copolymers and petroleum sulfonates are rather poorly biodegradable, and quats are already under attack. Questions will probably be raised about the aquatic toxicity of effluents containing these partly biodegraded surfactants. Of course, the volumes of these surfactants reaching the receiving waters are relatively small in comparison with the volumes of household surfactants, and EO/POs have already to some extent been replaced, but nevertheless it may be argued that if these materials can be detected, they must be reduced or replaced.

SURFACTANT	REPLACEMENT
APE	AE (several types)
EO/PO Copolymers	Modified AE
LABS	?
Petroleum Sulphonates	?
Quats	?

Figure 11 POSSIBLE REPLACEMENTS FOR INDUSTRIAL SURFACTANTS

Some industries may be forced to change not just an ingredient in their process but the actual process itself because of future environmental constraints. Industries which are large consumers of auxiliary chemicals and which generate a large volume of aqueous effluent (such as paper and textiles) are likely to be particularly affected. As we shall hear later in this conference, there is a particular need for nonionic surfactants which combine the properties of good wetting power, low foam generation and high biodegradability. In the surface-coatings industry, the use of volatile solvents is increasingly under attack on environmental and cost grounds, and this must offer opportunities for surfactants in water-based emulsion systems.

One industry which <u>is</u> more receptive to new surfactants is agrochemicals. New pesticides are always under development, bringing the opportunity for new formulations involving surfactants. It must be recognized, however, that the development of more active pesticides, together with environmental pressure to achieve more efficient spraying, will mean that the total consumption of surfactants is unlikely to increase much. On the other hand, the cost contribution of a surfactant in such a formulation is relatively minor, and this must present opportunities for alternative surfactants (perhaps as adjuvants, to increase the effectiveness of the toxicant).

The polymer industry is always interested in improved emulsifiers, particularly for surface coatings, even though "improvement" often means something they are able to use in smaller quantity!

The textile industry uses a very wide range of surfactants; some areas, notably synthetic fibers and fabric finishes, are progressive and are on the look out for new auxiliaries; other areas, e.g. cotton and wool processing, have to be more cost-conscious and tend to stick to their established procedures. With increasing competition from Asia, the prospects for the textile industry as a whole cannot be very bright. The petroleum industry offers more promise.

Although development of engines with better fuel efficiency will severely limit growth of petroleum consumption, environmental pressure to develop more efficient engines and reduce emissions will stimulate demand for lube oil and fuel additives to help achieve this higher performance.

In the paper industry, environmental pressure to increase recycling is stimulating the demand for surfactants in processes such as deinking.

When we are thinking about changing the surfactant used in an industrial process, we must realize that it is not as simple a matter as, say, reformulating a household detergent or shampoo.

Every industrial surfactant user is always interested in principle in having a better and/or cheaper product, but when it comes to the point, he may feel reluctant to depart from his established recipe. Key factors here are: how much will it cost to evaluate a new surfactant, and what are the possible consequences if something should go wrong?

I can illustrate this with some examples. A change in a surfactant used in an SBR emulsion polymerization may require product testing which includes extensive road trials of tires lasting up to 2 years. Evaluation of a surfactant as a foaming agent for manufacture of wallboard (gypsum board in the U.S.) calls for plant trials involving manufacture of hundreds of meters of board, which may be unsalable.

A change in the surfactant used in an agrochemical emulsion concentrate will entail registration of the new formulation, supported by very extensive testing including greenhouse trials.

Evaluation of a surfactant in concrete manufacture may involve preparation and mechanical testing of large beams which is outside the scope of a normal laboratory. The possible consequences of the failure of a large concrete structure (a bridge over a motorway, or a block of flats, for example), which could perhaps be attributed to a change of surfactant, instill caution in the surfactant users.

It should be clear from what I have said that in some industrial applications, the surfactant user will be reluctant to change his tried and trusted system until he is compelled to do so. Others are more receptive to new materials, or are under pressure to replace current surfactants, and these present real opportunities for the surfactant suppliers. Because of the extensive testing programs involved, it is essential to include the candidate new surfactants from the very beginning of the laboratory test programs.

In his introductory paper at the last conference here, David Karsa drew attention to the many different types of surfactants which have been prepared and reported in the literature but have never achieved commercial success. Often, I suspect, they were prepared for testing in one specific application, and if they failed, or there was no obvious economic manufacturing route available, they were put back on the shelf. We all know the obstacles, and the costs, in the way of commercializing a "new" chemical. In the area of surfactants, I believe it is only specialized high-value materials, which have some unique properties, that will have much chance of success as industrial surfactants if they are "new chemicals" within the meaning of the Act. Recent examples of successful novel surfactants are the oligomeric materials, and the various fluorine- and silicon-based products.

We can expect some "tweaking" of existing surfactant products, by which I mean juggling with substituents, lengths of polyethylene oxide chains, modified end-groups, etc., and further exploration of synergistic combinations of surfactants.

I said a few moments ago that the natural growth prospects for established surfactants in established industrial outlets are positive but strictly limited. The key to survival of the surfactant industry in this area is the development of substitution opportunities, based on improved cost/performance or better environmental compatibility.

c) Novel industrial applications for surfactants

When we come to try to think of new industrial applications for surfactants, it is difficult to come up with anything that has not been known, at least in the form of a patent application, for some years. The stumbling blocks, in most cases, are cost and the problems of evaluation which I have mentioned earlier. Thus, to take the most obvious example, the technology for micellar flooding for enhanced oil recovery has been available in principle for quite some time, and the potential surfactant requirement is enormous, but the process is uneconomic at present crude oil prices. Another example is the movement of coal/water slurries by pipeline and their use as fuel. This also is in abeyance at present because of the low cost of alternative fuels, but the day of this technology will surely come. Even more speculative are projects for reducing water evaporation from reservoirs or enhancing the fertility of desert soils by application of small amounts of surface-active materials.

We have summarized the outlook for industrial surfactants in Figure 12. Overall, we do not expect the growth in total surfactant consumption in industrial applications in W. Europe (or the U.S.) to exceed 2% per annum over the next five years. This of course is an average figure, made up of some sectors with above average growth (asphalt, paper, concrete) and some of below average growth (food, textiles) or even negative growth (leather). Although considerably smaller in volume terms than the industrial surfactant sector, the I&I cleaner sector is expected to have a rather brighter future, with an annual growth of around 3%. This reflects the growth of consumer and service industries (catering, laundry, office cleaning etc.) APE has already been phased out to a great extent, so major changes in surfactant consumption patterns are not foreseen in this sector. Because of substitution of materials such as APE, EO/PO copolymers, and perhaps soap or

1. Total consumption + <2%/yr
Positive areas: I&I Cleaners, Asphalt, Paper, Concrete
Slow areas: Food, Textiles, Leather

2. SFA substitution for environmental or performance reasons
(Cost of reformulation = inhibiting factor)

3. Needs: new processes, applications, materials recycling, pollution reduction

Figure 12 THE OUTLOOK

even LABS, for environmental or performance reasons, the growth prospects for individual surfactants are better than this. AE and various types of modified AE should be the main beneficiaries from the enforced reduction in APE consumption.

The other aspect of the future is the need for some new processes and applications involving surfactants.

The increased attention being given to materials recycling and to the reduction of pollution may hold some promise for surfactants.

Concluding Remarks

It is very difficult to estimate the size of the industrial surfactant market. Statistics are lacking, and there are problems of definition. Most industries use surfactants at some stage, although they may be called process aids or auxiliaries.

In the U.S., industrial uses account for about half the total surfactant consumption, amounting to more than 1 millon tons (without soap). In West Europe, industrial uses account for about one-third of the total surfactant consumption.

The growth prospects for industrial surfactants vary from one industry to another. Total surfactant consumption is not expected to increase at more than about 2 percent per annum, as the economies of the Western countries come out of the present recession. Environmental pressures will become more important in those applications where there is an aqueous effluent, and surfactants such as APE will increasingly be substituted by alternative surfactants with better environmental properties. This provides good growth opportunities for individual surfactants.

Environmental issues have not yet become as dominant as they are in the household detergents area, but nevertheless interest may be expected increasingly to focus on aqueous effluents, and the biodegradability and aquatic toxicity of any surfactants they contain.

Many of the industrial processes which use surfactants are long-established and there is a real need for some new processes or applications to revitalize the industry. These may not necessarily involve totally new surfactants. There are still outlets for established materials, alone or in combinations. It should also be remembered that many interesting materials have been described in the textbooks but never commercialized, and are now waiting to be re-discovered. Applications work needs to be done, particularly in those areas where performance can be assessed on the basis of laboratory screening.

When considering possible changes in surfactant use patterns, a key factor is the amount of testing (and associated financial risk) involved on the part of the surfactant users.

Finally, we should remember that West Europe is some way behind the U.S. in its industrial uses of surfactants. This must mean opportunities - for you!

References

1. Surfactants in Industrial Applications. Western Europe Forecast to 1995. Colin A. Houston & Associates Inc. 1985.

2. Industrial Applications of Surfactants. North American Forecast to 2000. Colin A. Houston & Associates Inc., 1991.

3. Chemical Week, Jan. 30, 1991, 38

4. Chemicals in Petroleum Exploration and Production III - North American Report and Forecast to 2000. Colin A. Houston & Associates Inc., 1991.

Innovation

Surfactants and the Environment — Some Recent Developments

[1]G. Bognolo, [2]G.R. John and [2]J.G. Evans

[1]ICI SURFACTANTS, EVERSLAAN 45, B-3078 EVERBERG, BELGIUM
[2]ICI SURFACTANTS, P.O. BOX 90, WILTON CENTRE, MIDDLESBROUGH, CLEVELAND TS6 8JE

The chemical industry as a whole is very aware of and responsive to the environmental issue. Given the multitude of products used in such a very large and diverse number of applications, surfactant producers, in particular, find themselves faced with increasing demands for products that are seen to be "non-harmful" to the environment, both during preparation and in their end-use.

Surfactant biodegradability is a major issue (compliance with EEC directives and concerns over residues from biodegradation) and market driven initiatives are influencing already the types and usage of surfactants across many industrial applications. In addition to this, surfactant producers have to respond to changes related to the use of solvents (safety, VOC requirements) and the use of chlorinated hydrocarbons (Montreal Declaration, COSHH). Even product packaging, waste disposal and associated effluent management are becoming very important considerations. All are part of the environmental issue.

Further these pressures on surfactant producers are compounded by demands for high performance surfactants, increasing rationalisation across industry sectors and the "quality" culture. The need for new products, new formulations and end-user processes has never been greater. As a consequence it is inevitable that some traditional surfactant products will come under threat, whilst others attain greater acceptability and importance. It is clear though, that surfactant producers who are best able to respond positively to the demands of the new situation will find many opportunities.

Significant changes already facing surfactant producers include:

(i) The growing market pressure to reduce the usage of alkylphenol ethoxylates and replace them with suitable alternatives.

(ii) The replacement of "hard" EO/PO block copolymers with more readily biodegradable copolymers, and

(iii) The progressive shift from solvent-based formulations and processes to aqueous-based systems.

These changes are taking place apace across many industrial applications and reformulation activities are under way in most sectors. In the industrials surfactants area, resin emulsification as a method of reducing the VOC in adhesives, paints and inks shows great promise. For industrial fluids advances in water-based metal working fluids and lubricants are becoming possible through the availability of more effective emulsifiers, while in agrochemical formulations the development of more effective dispersants is meeting the need for lower level dispersant/lower solvent containing pesticide formulations. The impact of change is especially significant in the area of industrial cleaning, the so-called Industrial and Institutional (I&I) cleaning market, where the products, their formulation and processes of use may impact greatly with the environment. New product developments in alcohol ethoxylates, alkylpolysaccharides, ether carboxylates, end-capped ethoxylates, etc, as well as modifications to existing product types, are seen as a way of meeting the new product opportunities opposite changing industry demands.

ICI Surfactants is committed to the pursuit of environmentally acceptable products, their preparation and end-use, across all industry sectors. Development work in each of the above mentioned industrial applications is in progress and ICI Surfactants is working with formulators to assist them through this time of great change.

In view of the importance of the I&I cleaning market to surfactant producers some recent developments are reviewed, together with a brief description of the area.

The I&I cleaning market itself is large fragmented and diverse. For Europe it is estimated (1990)at around 170,000 tons and can be segmented into six application areas:-

hard surface cleaners	57,000 tons
laundry/dry cleaning	53,000 tons
dishwashing	27,000 tons
food industry processing	13,000 tons
metal cleaning	13,000 tons
other.	9,000 tons

The hard surface cleaners are primarily janitorial applications and the products used are very similar to household equivalents. Likewise, laundry/dry cleaning products tend also to be based on household equivalents, but here the environmental issues around the use of chlorinated solvents in dry cleaning are driving the market towards aqueous-based cleaning systems. In contrast because the scale and design of industrial dishwashers are considerably different from household equivalents, the products used are different. Operating conditions are more extreme and so the performance requirements of the formulations are much higher. Food industry processing is a very wide area covering dairy cleaning, brewery cleaning through to cleaning of food production equipment, so while the requirements may be diverse, the standards of cleaning are very high. Metal cleaning is undergoing major change. Chlorinated solvents have been the main cleaning medium for many years, but as in dry cleaning, the environmental concerns are accelerating the move to aqueous-based systems where possible.

It is clear, then, that in addition to major factors such as hygiene, rationalisation across the cleaning industry and the changing economic climate, the environmental issue has become the major influence for change. Historically the I&I cleaning market has been slow to respond to environmental pressures, unlike the household cleaning market which responded several years ago, but with increasing legislation and consumer pressures, this market is following the household market very rapidly.

The use of many standard surfactants is already decreasing as a result of the formulators turning to environmentally more acceptable products. The typical usage pattern for surfactants going into the I&I cleaning market a year or so ago would have been:-

alkylphenol ethoxylates	20%
alcohol ethoxylates	21%
EO/PO copolymers	8%
linear alkylbenzene sulphates	21%
alcohol (ether) sulphates	6%
phosphate esters	9%
others	15%

The most significant changes taking place concern the trend away from alkylphenol ethoxylates and increasing importance of alcohol ethoxylates, as well as the replacement of "hard" EO/PO block copolymers by more acceptable copolymer products.

The major issue is to meet environmental requirements and at the same time give the required performance. Often performance may be the key criterion that cannot be met by a single change and modification of formulation and conditions or processes of use may have to be considered. This is most obvious, for example, in the move from chlorinated solvents to aqueous, or semi-aqueous cleaning systems in metal degreasing. Furthermore, the optimum balance of properties for surfactants containing formulations may involve quite contradictory requirements, such as:-

biodegradability
foam control/generation
effective cleaning
alkaline stability
wetting/detergency
rinsing/drying
effluent management

ICI Surfactants produce and supply products as raw materials for all areas of the I&I cleaning market and is committed to developing and extending its broad portfolio of products to meet the new demands, fully utilising its expertise in surfactant chemistry and its understanding of formulation and application requirements. This may involve product modifications, changes in product types, and also new product developments, including "designer" surfactant products where appropriate. It is natural that changes in surfactant raw materials and their formulation will be accommodated more readily by this industry sector than the necessity for substantially new processes or operating conditions. This too, would avoid the end-users having the burden of changes in equipment and associated expenditure. With the move away from traditional surfactants set to continue, it has to be recognised that solutions to emerging problem areas will be found more readily in some areas than in others.

Three recent developments by ICI Surfactants in key areas of the I&I cleaning market are outlined below:-

(i) Nonylphenol Ethoxylate Replacement

Nonylphenol ethoxylates are very versatile products that have found extensive use over many years. They have been employed in simple formulations (for example, hard surface cleaners) through to more complex emulsion based systems (for example, car-care microemulsions). Their replacement is not straight forward on a 1:1 basis, but requires a degree of reformulation effort, particularly when the surfactant is the primary emulsifier for a hydrocarbon solvent. It is not surprising then, that no single replacement type has been identified. However, it has been demonstrated that in many instances, nonylphenol ethoxylate-based systems may be satisfactorily replaced with those based on currently available alcohol ethoxylates. With the homologue distribution of the alcohol ethoxylates being considerably broader than that of the nonylphenol ethoxylates, together with the free alcohol content of the former products which can lead under certain circumstances to end-use odour problems, careful selection of one or more can avoid such difficulties.

ICI Surfactants has several ranges of alcohol ethoxylates currently available and has built up substantial experience in this area which is available to assist the needs of formulators in solving reformulation problems. A formulation guidelines brochure is available.

(ii) Automatic Dish Wash Rinse Aids

With legislation in this area - derogations do not apply - there is the clear need to replace traditionally used surfactants in this area. These are primarily based on A-B-A block EO/PO copolymers and the higher molecular weight alcohol alkoxylates.

Product development work carried out by ICI Surfactants has resulted in the availability of a wide range of alcohol alkoxylates which meet the EEC directive and the performance requirements of the industry. ICI Surfactants has brochures available specifically for this area, which cover products, application and formulation guidelines.

(iii) Alkaline Cleaning

Food industry cleaning and metal degreasing are two areas where alkaline cleaning is commonly used. In these situations the surfactant must withstand the conditions of use for extended periods. The key requirements are specifically:-

alkaline stability
wetting
low foam

in addition to the important requirement of biodegradability.

ICI Surfactants has developed a product that achieves alkaline stability through replacement of a terminal - OH group. The other requirements are also fully met, such that all indications are that this product satisfies both performance requirements of the industry and the biodegradability directive. The product should be commercially available very soon.

In conclusion, ICI Surfactants is working to develop products, technology and expertise to meet the challenge of the changes brought about particularly as a direct result of the environmental issue. The above three examples illustrate where ICI Surfactants is able to offer solutions to major problems in the I&I cleaning market. Current activities, of course, cover the breadth of industrial applications in the same way, and reflect ICI Surfactants response to the challenge of meeting the new requirements of "surfactants and the environment".

Synthesis of Polymerizable Surfactant and its Application to Emulsion Polymerization

Kinya Yokota, Akinobu Ichihara and Hitoshi Shin'ike

APPLICATION LABO, DAI-ICHI KOGYO SEIYAKU CO. LTD., 55 NISHI-SHICHIJO HIGASHI-KUBOCHO, SHIMOGYO-KU, KYOTO 600, JAPAN

As polymerizable surfactants, 1-nonylphenoxy 2-polyoxyethylene 3-allyloxypropane, its sulphate and its phosphate were synthesized. 1-polyoxyethylene 2-propenyl 4-nonylphenol ether, its sulphate and its phosphate were also synthesized. The yields of those polymerizable surfactants are more than 95% and their syntheses are available due to the technique of ethoxylation, sulphation and phosphation.

These polymerizable surfactants were applied to emulsifier of emulsion polymerization of ethylacrylate and butylacrylate/styrene.

The result showed that their polymerization stabilities were the same level as that of surfactants commonly used and great improvements were seen in low foaming property of emulsion and water-resisting property of polymer film. In other words, a polymerizable surfactant is very effective as an emulsifier for emulsion polymerization.

1. Introduction

Synthetic resin emulsions made by the emulsion polymerization method are used as materials of adhesion, tacky adhesion, coating and so on. They are also used in the various fields including paints, textile, construction and paper.

However, it is generally said that such emulsions have inferior properties compared to solvent-type polymers because of bad effects of used surfactants.[1)−3)]

In order to eliminate such bad effects, polymerizable surfactants are proposed and some of them have been reported, but there is much more room for improvement and future developments are expected.[4)−5)]

We synthesized various kinds of polymerizable surfactants having polyoxyethylene chains whose polymerizable double bond is an allyl group or propenyl group, and studied estimation to emulsion polymerization.

2. Experimental

2-1 Synthesis of Polymerizable Surfactants

1) Synthesis of materials of hydrophobic group

1-nonylphenoxy 2-hydroxy 3-allyloxypropane 1-nonylphenoxy 2-hydroxy 3-allyloxypropane was synthesized by reaction of nonylphenol with allylglycidyl ether at 100°C for 3 hours using amine catalyst, non-reacting allylglycidyl ether and amine catalyst were removed by reducing pressure at 120°C by 10mmHg, and 1-nonylphenoxy 2-hydroxy 3-allyloxypropane was acquired. Products were confirmed because the peak level of allyl group was shown to be 5.2 ppm and 6.0 ppm by H-NMR.

$H_{19}C_9-C_6H_4-OH + CH_2(-O-)CHCH_2OCH_2CH=CH_2 \xrightarrow{\text{Amine}} H_{19}C_9-C_6H_4-OCH_2CH(OH)CH_2OCH_2CH=CH_2$

Scheme 1

2-allyl 4-nonylphenol Allylnonylphenyl ether was synthesized by reaction of nonylphenol with allyl chloride at 40°C for 5 hours using NaOH catalyst. Purification of allylnonylphenyl ether was carried out by leaving it quiet until separation of water was observed. Separated water in the lower part was removed and the upper part was rinsed by adding water. After this process was continued, allylnonylphenyl ether was acquired. Then 2-allyl 4-nonylphenol was obtained by heating allylnonylphenyl ether at 200°C for 5 hours by Claisen rearrangement. The presence of the allyl group was confirmed by its peaks at 3.3 ppm, 5.2 ppm and 6.0 ppm by using H-NMR.

$H_{19}C_9-C_6H_4-OH + ClCH_2CH=CH_2 \xrightarrow{\text{NaOH}} H_{19}C_9-C_6H_4-OCH_2CH=CH_2$

$\xrightarrow[\text{200 °C}]{\text{Claisen Rearrangement}} H_{19}C_9-C_6H_3(CH_2CH=CH_2)-OH$

Scheme 2

2) Synthesis of Polymerizable Surfactants

Nonionic Surfactants Nonionic polymerizable surfactants which contain polymerizable double bonds were synthesized by well-known ethoxylation of 1-nonylphenoxy 2-hydroxy 3-allyloxypropane or 2-allyl 4-nonylphenol as below. Allyl group was observed to convert to propenyl group in the process of adding ethylene oxide to 2-allyl 4-nonylphenol.

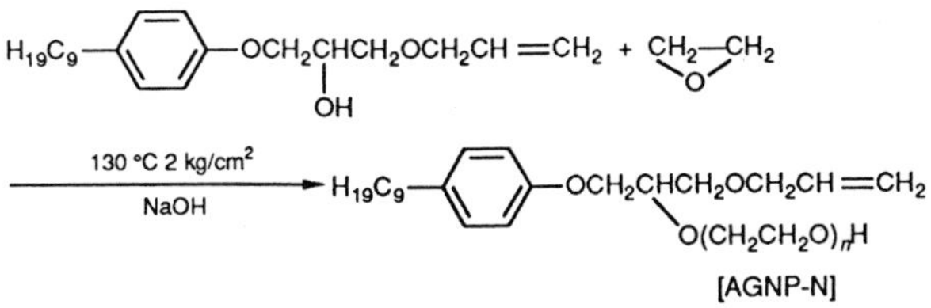

$H_{19}C_9-C_6H_4-OCH_2CH(OH)CH_2OCH_2CH{=}CH_2 + CH_2-CH_2$ (O)

$\xrightarrow[\text{NaOH}]{130\ °C\ 2\ kg/cm^2} H_{19}C_9-C_6H_4-OCH_2CH(O(CH_2CH_2O)_nH)CH_2OCH_2CH{=}CH_2$

[AGNP-N]

1-nonylphenoxy 2-polyoxyethylene 3-allyloxypropane

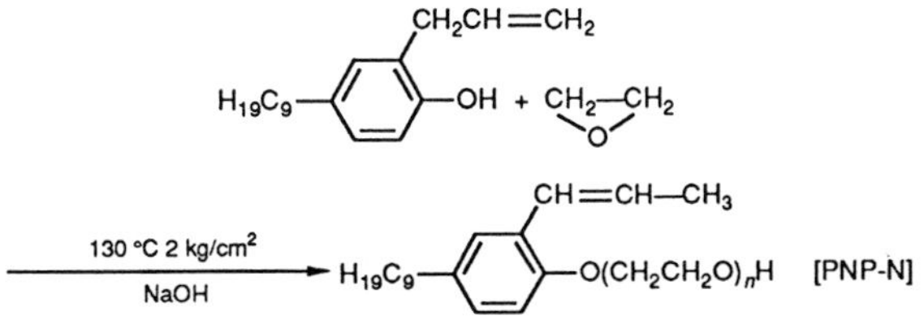

$H_{19}C_9-C_6H_3(CH_2CH{=}CH_2)-OH + CH_2-CH_2$ (O)

$\xrightarrow[\text{NaOH}]{130\ °C\ 2\ kg/cm^2} H_{19}C_9-C_6H_3(CH{=}CH-CH_3)-O(CH_2CH_2O)_nH$ [PNP-N]

1-polyoxyethylene 2-propenyl 4-nonylphenyl ether

Scheme 3

Sulphate-type Anionic Surfactant Sulphate-type anionic polymerizable surfactant was synthesized by sulphation with sulphamic acid of nonionic surfactant having polymerizable double bond in the above process as below.

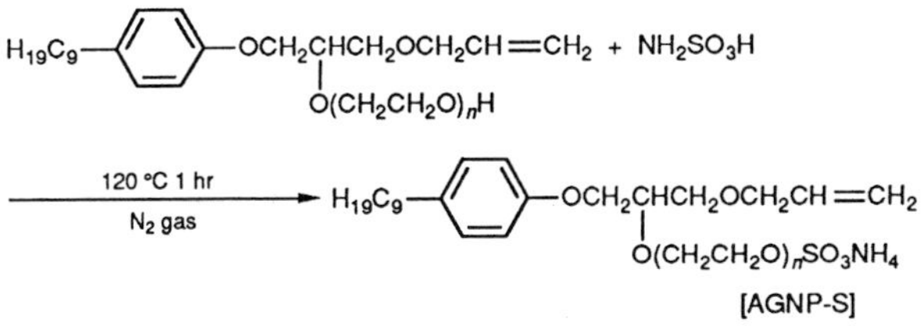

$H_{19}C_9-C_6H_4-OCH_2CH(O(CH_2CH_2O)_nH)CH_2OCH_2CH{=}CH_2 + NH_2SO_3H$

$\xrightarrow[N_2\ \text{gas}]{120\ °C\ 1\ hr} H_{19}C_9-C_6H_4-OCH_2CH(O(CH_2CH_2O)_nSO_3NH_4)CH_2OCH_2CH{=}CH_2$

[AGNP-S]

1-nonylphenoxy 2-polyoxyethylene 3-allyloxypropane ammonium sulphate

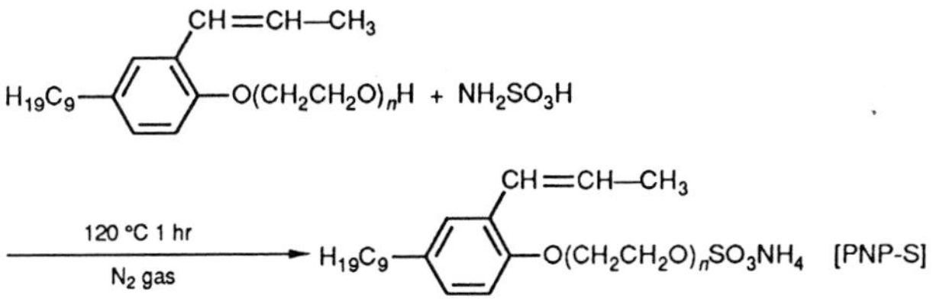

$H_{19}C_9-C_6H_3(CH{=}CH-CH_3)-O(CH_2CH_2O)_nH + NH_2SO_3H$

$\xrightarrow[N_2\ \text{gas}]{120\ °C\ 1\ hr} H_{19}C_9-C_6H_3(CH{=}CH-CH_3)-O(CH_2CH_2O)_nSO_3NH_4$ [PNP-S]

1-polyoxyethylene 2-propenyl 4-nonylphenyl ammonium sulphate

Scheme 4

Phosphate-type Anionic Surfactant Phosphate-type anionic surfactant was synthesized by phosphation with phosphoric anhydride of nonionic surfactant having polymerizable double bond in the above process as below.

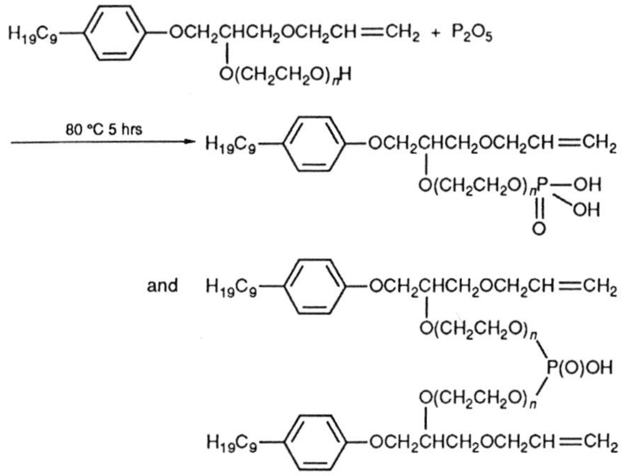

$H_{19}C_9-C_6H_4-OCH_2CH(O(CH_2CH_2O)_nH)CH_2OCH_2CH{=}CH_2 + P_2O_5$

$\xrightarrow{80\ °C\ 5\ hrs}$ $H_{19}C_9-C_6H_4-OCH_2CH(O(CH_2CH_2O)_nP(=O)(OH)_2)CH_2OCH_2CH{=}CH_2$

and $[H_{19}C_9-C_6H_4-OCH_2CH(CH_2OCH_2CH{=}CH_2)O(CH_2CH_2O)_n]_2P(O)OH$

1-nonylphenoxy 2-polyoxyethylene 3-allyloxy mono- or diphosphate

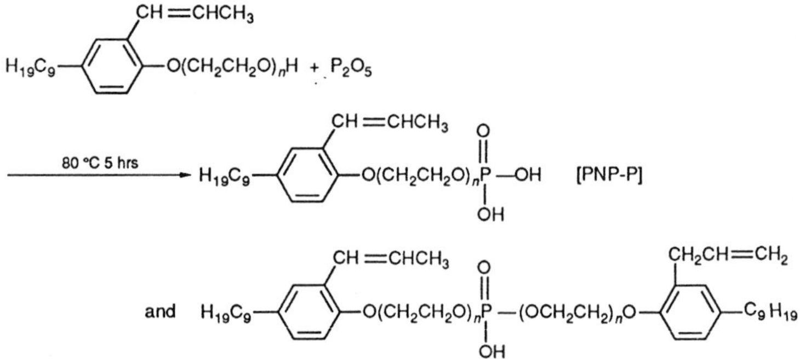

$H_{19}C_9-C_6H_3(CH{=}CHCH_3)-O(CH_2CH_2O)_nH + P_2O_5$

$\xrightarrow{80\ °C\ 5\ hrs}$ $H_{19}C_9-C_6H_3(CH{=}CHCH_3)-O(CH_2CH_2O)_nP(=O)(OH)-OH$ [PNP-P]

and $H_{19}C_9-C_6H_3(CH{=}CHCH_3)-O(CH_2CH_2O)_nP(=O)(OH)-(OCH_2CH_2)_nO-C_6H_3(CH_2CH{=}CH_2)-C_9H_{19}$

1-polyoxyethylene 2-propenyl 4-nonylphenol mono- or diphosphate

Scheme 5

2-2 Emulsion Polymerization Test

1) Sample

Monomers were used, after ethylacrylate, butylacrylate and styrene were rinsed firstly with sodium bisulphite solution, secondly with 5% sodium hydroxide solution, and thirdly with 20% sodium chloride solution, then dehydrated by sodium sulphate, and given vacuum-distillation treatment. As for ammonium persulphate, guaranteed reagent made by Nakarai Chemicals, Ltd. was used, and common ion-exchanged water was adopted. As for polymerizable surfactants, the above mentioned compounds were used for comparion with them, conventional surfactants were adopted, such as HITENOL N-08 (polyoxyethylene (n=6) nonlylphenol ammonium sulphate), NEOGEN R (sodium dodecylbenzene sulphonate (hard type)), NOIGEN EA170 (polyoxyethylene (n=21) nonylphenyl ether), PLYSURF A212E (polyoxyethylene (n=8.6) nonylphenyl ether phosphate) made by DAI-ICHI KOGYO SEIYAKU CO., LTD.

2) Emulsion Polymerization Test

A 500-mL separable flask with a condenser, a thermometer, and a dropping funnel was used for testing emulsion polymerization.

Firstly, 273.6g of ion-exchanged water was put into the flask, the temperature was raised to 80°C, Nitrogen gas was put into the water for 10 min., and oxygen in the water was removed.

After 6.0g of surfactant was dissolved in 200g of ethylacrylate, 20g was extracted from the liquid, and added to the flask. Then the temperature was adjusted at 80°C, and polymerization initiator solution dissolved 0.4g of ammonium persulphate into 20g of water was added into the flask. After heat was generated in the content of the flask and emulsion conditions were converted, starting of polymerization was observed. Then the rest of the blended liquid of monomer and surfactant were dropped constantly into the flask for 3 hours keeping the inside temperature at 80 ± 1°C.

After dropping was finished, it was left quiet for 1 hour at 80 ± 1°C, cooled, filtered through a 150 mesh stainless net and emulsion was extracted

The recipe of butylacrylate/styrene/acrylic acid=99/99/2 was the same as the method of emulsion polymerization of ethylacrylate.

Estimation

Coagulation : Following polymerization, the emulsion was filtered through a 150 mesh stainless net, the residue was dried on the net at 105°C, its weight was measured, and its ratio against the whole monomers was indicated by percentage.

Non-Volatile Matter : 1g of emulsion was dried at 105°C for 2 hours, its weight was measured, and its ratio against emulsion was indicated by percentage.

Viscosity : Viscosity was measured by a BM-type viscometer at 30 rpm and at 25°C.

Particle Size : Particle size was measured by the method of dynamic light scattering measurement using DLS-700 light scattering spectrophotometer made by Otsuka Electronics.

Surface Tension : Surface tension of emulsion was measured by Wilhelmytype surface tensionmeter at 25°C.

Emulsion Foaming : Emulsion was diluted with the same amount of water, 50mL was extracted from it, and put into a 100mL Nessler tube. Then it was revolved at 180° thirty times, left quiet, and the quantity of foaming was measured.

Chemical Stability : 10 mL of $CaCl_2$ solution of various concentrations was added to 10 mL of emulsion with stirring, and the concentration of $CaCl_2$ was indicated by mol/l at coagulation of polymers.

Mechanical Stability : After 50g of emulsion was adjusted with ammonium hydroxide at 9 ± 1 pH, mechanical stability was measured by a Marontype tester. (Pressure : 10 kg, Speed of Agitation : 1,000rpm, 5 min., 80 Mesh On)

Water-Resistance : A latex film of 0.5 mm in thickness was made on a slideglass, and it was left as it was for 18-20 hours. Then it was immersed in the water, and it was observed that how long it took for a letter of 4.5 point through the film to become obscured.

Contact Angle of the Film : Latex film was made on a slideglass, and contact angle of a drop of the water was measured.

3. Result and Discussion

3-1 Synthesis of Polymerizable Surfactant

(1) Polymerizable Surfactant Having 1-Nonylphenoxy 2-Hydroxy 3-Allyloxypropane as Hydrophobic Group

The yield of synthesis of hydrophobic group is 98-99%. In the process of synthesizing nonionic sulphate-type and phosphate-type surfactants, no disappearance of allyl group as polymerizable double bond was observed.

(2) Polymerizable Surfactant having 2-Allyl 4-Nonylphenol as Hydrophobic Group

The yield of synthesis of hydrophobic group is 95-96%. It has been already known that allyl group is converted into propenyl group when ethylene oxide is added to it under the existence of NaOH.[6]

In the process of synthesizing sulphate-type and phosphate-type surfactants, no disappearance of propenyl group as polymerizable double bond was observed.

Test Results of Synthesized Polymerizable Surfactants

Table-1 Nonionic Polymerizable Surfactant

Product	OH V (KOH mg/g)	Moisture (%)	Double Bond(mg eq/g) Theorètical/Actual Value
AGNP-N (n=20)	47.1	0.04	0.83 / 0.81
" (n=50)	23.4	0.11	0.40 / 0.39
PNP-N (n=20)	49.6	0.05	0.88 / 0.84
" (n=50)	23.8	0.10	0.41 / 0.39

Table-2 Sulphate-type Anionic Polymerizable Surfactant

Product	Conversion of Sulphation	pH	Volatile Matter (%)	Double Bond(mg eq/g) Theoretical/Actual Value
AGNP-S(n=2)	94	7.3	0.4	1.99/1.95
" (n=6)	95	7.5	0.3	1.47/1.43
" (n=10)	96	7.4	0.3	1.17/1.14
" (n=20)	95	7.7	0.4	0.77/0.75
PNP-S (n=10)	96	7.6	0.3	1.25/1.19
" (n=20)	96	7.7	0.6	0.81/0.76

Table-3 Phosphate-type Anionic Polymerizable Surfactant

Product		Monophosphate/ Diphosphate (Molar Ratio)	Acid Value (KOH mg/g)	Moisture (%)	Double Bond (mg eq/g) Theoretical/ Actual Value
AGNP-P	(n=2)	54/46	115	0.12	2.21/2.16
"	(n=6)	58/42	85	0.08	1.60/1.55
"	(n=10)	62/38	65	0.11	1.24/1.21
PNP-P	(n=10)	60/40	71	0.10	1.34/1.27

3-2 Physical Properties of Polymerizable Surfactants

The below Table-4, 5, 6, Figure-1 and 2 show the results of surface tension measurement, foaming property and emulsifying property of each synthesized polymerizable surfactant. As a result,
it was found that polymerizable surfactants had the same properties of conventional surfactants. It means that combining double bond with hydrophobic group doesn't cause any bad effects on polymerizable surfactant.

Table-4 Surface Tension
(0.1% soln, Wilhelmy's Method, 25°C)

Product	Surface Tension (dyne/cm)
AGNP-N (n=20)	39.9
PNP-N (n=20)	38.3
AGNP-S (n=10)	41.2
PNP-S (n=10)	40.5
AGNP-P (n=10)	37.6
PNP-P (n=10)	37.8
NOIGEN EA170	39.9
HITENOL N-08	40.0
PLYSURF A212E	38.1

Table-5 Foaming Property
(0.1% soln, Ross Miles' Method, 25°C)

Product	Height of Foam (mm) Immediately after	After 5 min.
AGNP-N (n=20)	154	123
PNP-N (n=10)	150	140
AGNP-S (n=10)	125	107
PNP-S (n=10)	190	170
NOIGEN EA170	182	145
HITENOL N-08	167	155

Table-6 Emulsifying Property

Product	Amount of Separated Water After 3 hours (ml)
AGNP-N	0
PNP-N	0
AGNP-S	1
PNP-S	1
NOIGEN EA170	1
HITENOL N-08	0

Emulsifying Prescription : Surfactant 1
Water 49
Butylacrylate 50

Universal Homogenizer : 2,000 rpm, Blending for 10 min., Leaving the mixture quiet in a 100 mL cylinder with a cap.

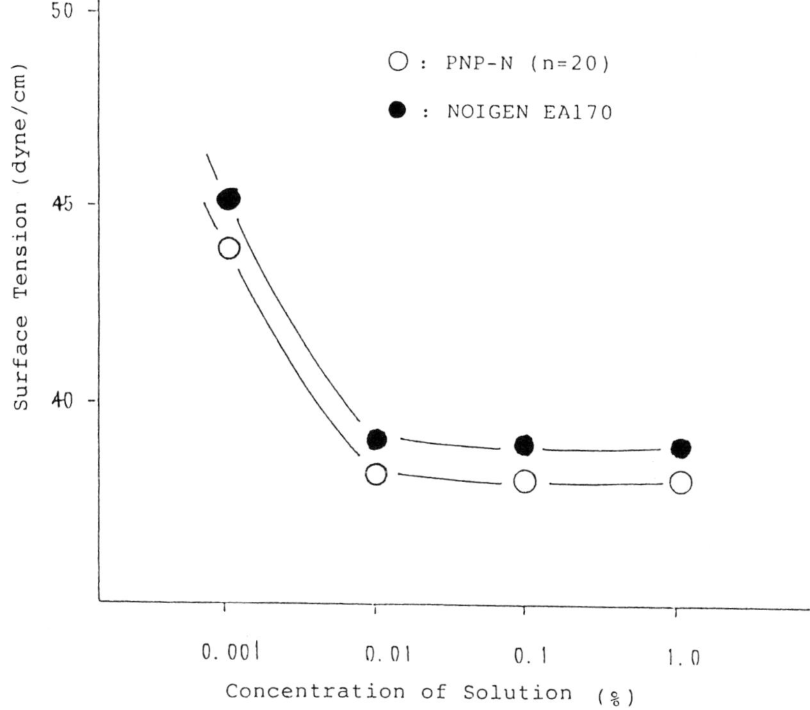

Fig. 1 Measuring Method of Surface Tension in Surfactant Solution: Wilhelmy-type Surface Tensiometer

Temperature : 25°C

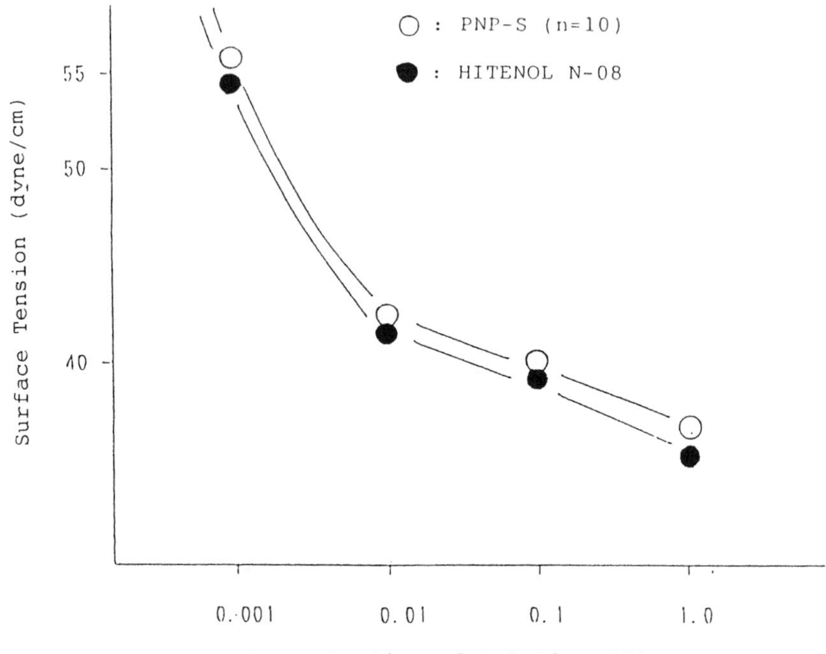

Fig. 2 Measuring Method of Surface Tension in Surfactant Solution: Wilhelmy-type Surface Tensiometer

Temperature : 25°C

3-3 Copolymerizability of Polymerizable Surfactant

(1) Copolymerizability of Nonionic Polymerizable Surfactant

40g of monomer (ethylacrylate or butylacrylate), 60g of benzene, 0.2g of azo-bis-iso butyronitrile and 0.8g of nonionic polymerizable surfactant were blended, they were polymerized at 60-65°C for 4 hours, and polymer was acquired. Its unpolymerized nonionic polymerizable surfactant was determined by GPC.

Table-7 Conversion Ratio of Nonionic Polymerizable Surfactant to Copolymer

Monomer	Polymerizable Surfactant	Conversion to Copolymer (%)
Ethylacrylate	AGNP-S (n=20)	76.6
"	PNP-N (n=20)	79.7
Butylacrylate	AGNP-N (n=20)	70.2
"	PNP-N (n=20)	76.7

(2) Copolymerizability of sulphate-type Polymerizable Surfactant

Polymer emulsion was made by using 3% of AGNP-S (n=10) or PNP-S (n=10) against monomer. Then the emulsion was dissolved into/in the solvent, which is dissolving PEA into dioxane and dissolving PBA/Pst into THF, polymer was precipiated in the water again, and unpolymerized surfactant dissolved in the water layer was determined by the Epton's method.

Table-8 Conversion Ratio of Sulphate-type Polymerizable Surfactant of Copolymer

Monomer	Sulphate-type Polymerizable Surfactant	Conversion to Copolymer(%)
Ethylacrylate	AGNP-S (10)	84.3
"	PNP-S (10)	85.9
Butylacrylate	AGNP-S (10)	82.2
"	PNP-S (10)	83.0

3-4 Emulsion Polymerization Test

The results of emulsion polymerization test of ethylacrylate and butylacrylate/styrene are shown in Table-9 and10.

(1) Polymerization Stability

Excellent stability, similar to conventional surfactants, was observed in all of the nonionic, sulphate-type or phosphate-type of polymerizable surfactants. In general, when monomer polymerizes, polymerizable surfactant also copolymerizes and it is taken inside polymer. So the quantity of surfactant on the surface of particles is short. As a result, polymerization stability deteriorates in using polymerizable surfactant. It was found that a hydrophilic group creates a good effect on the surface of particles because polymerizable double bond is located on the hydrophobic group side and effective stabilizing function is made by polyethylene oxide chain.

As for the linkage between moles of ethylene oxide and polymerization stability, in the case of nonionic polymerizable surfactant, the more the degree of ethoxylation increased, the better polymerization stability became. (Fig.3) In the case of sulphate-type polymerization, anionic polymerizabile surfactant, no remarkable difference was observed.

Table-9 Result of Emulsion Polymerization of Ethylacrylate

Emulsifier	Coagulation (%)	Non-Volatile Matter (%)	Viscosity (cP)	Particle size (μ)	Surface Tension (dyne/cm)	Foaming Property of Emulsion(ml) Immediately	After1min.	After5min.	Chemical Stability $CaCl_2$ (mol/ℓ)	Mechanical Stability (%)	Water Resistance (hr)	Contact Angle of Film (°)
AGNP-N (n=20)	4.37	38.4	10	0.81	39.8	40	38	25	over 2.0	0.00	over 168	63
PNP-N (n=20)	4.41	40.3	10	0.78	39.9	25	20	16	over 2.0	0.00	over 168	81
AGNP-S (n=10)	0.11	40.5	330	0.10	53.0	25	4	0	0.5	0.35	over 168	81
PNP-S (n=10)	0.11	40.6	97	0.10	48.9	21	0	0	1.0	0.00	over 168	83
AGNP-P (n=10)	0.18	38.4	16	0.16	46.1	19	0	0	1.0	0.10	over 168	78
PNP-P (n=10)	0.22	38.7	15	0.14	45.8	17	0	0	1.0	0.00	over 168	80
NOIGEN EA170	4.24	38.9	7	0.70	39.5	42	26	9	over 2.0	0.00	2	9
HITENOL N-08	0.04	40.3	24	0.09	41.3	39	37	32	1.0	0.00	4	12
PLYSURF A212E	0.26	38.3	18	0.14	40.6	38	32	25	1.0	0.10	3	8
NEOGEN R	0.16	39.3	21	0.07	38.2	56	56	53	0.03	1.12	3	6
AGNP-S(n=10)/AGNP-N(n=20)	0.01	40.9	122	0.12	51.4	25	2	0	over 2.0	0.09	over 168	78
PNP-S(n=10)/PNP-N(n=20)	1.16	40.5	86	0.12	51.4	25	9	5	over 2.0	0.64	over 168	82
HITENOL N-08/NOIGEN EA170	0.22	39.1	28	0.10	43.8	28	22	11	over 2.0	0.01	4	10

Mixing Ratio of Surfactant : anion/nonion=3/7

Table-10 Result of Emulsion Polymerization of Butylacrylate/Styrene

Emulsifier	Coagulation (%)	Non-Volatile Matter (%)	Viscosity (cP)	Particle size (μ)	Surface Tension (dyne/cm)	Foaming Property of Emulsion (ml) Immediately	After 1 min.	After 5 min.	Chemical Stability $CaCl_2$ (mol/ℓ)	Mechanical Stability (%)	Water Resistance (hr)	Contact Angle of Film (°)
AGNP-S (n=10)	0.02	39.4	505	0.08	61.2	19	5	0	0.3	1.2	over 168	74
PNP-S (n=10)	0.02	38.1	305	0.07	60.0	18	1	0	0.3	1.4	over 168	80
HITENOL N-08	0.11	39.7	151	0.07	53.1	35	22	6	0.3	3.4	10	below 5
NEOGEN R	0.16	38.6	413	0.08	47.7	47	38	32	0.03	1.8	5	below 5
AGNP-S(n=10)/AGNP-N(n=20)	0.54	39.5	47	0.10	57.2	22	0	0	1.0	0.3	over 168	82
PNP-S(n=10)/PNP-S(n=20)	1.63	38.1	59	0.09	56.1	18	1	0	0.03	0.7	over 168	84
HITENOL N-08/NOIGEN EA170	0.30	37.4	50	0.09	51.6	25	10	4	1.0	1.6	24	12

Mixing Ratio of Surfactant : anion/nonion=3/7

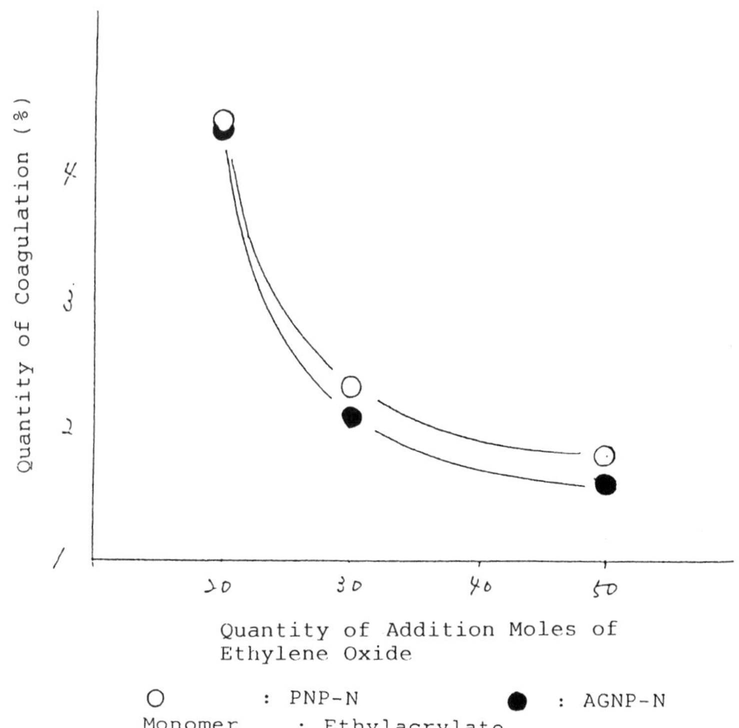

Fig. 3 Quantity of Ethylene Oxide Addition Moles in Polymerizable Surfactant and Quantity of Coagulation

(2) Foaming Property of Emulsions

Less foaming and excellent defoaming properties were observed in emulsions using AGNP-S (n=10), or PNP-S (n=10) as sulphate-type anion polymerizable surfactant. As for emulsions using nonionic polymerizable surfactants, less foam was observed, but its defoaming property was inferior to conventional nonionic surfactant. Never the less, if it was used with anionic polymerizable surfactant, defoaming property was excellent. These results are shown in Fig.4 and 5.

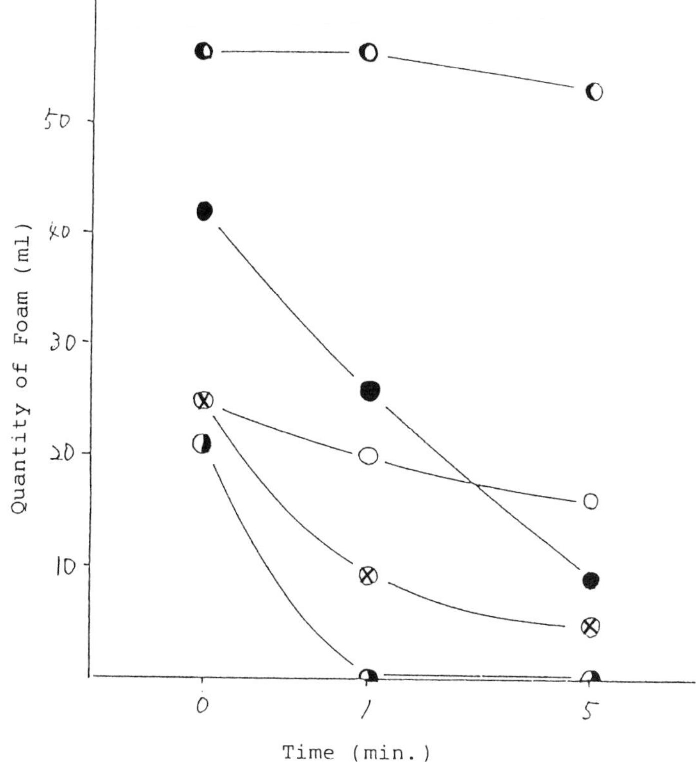

Fig. 4 Foaming Property of Ethylacrylate Emulsion

○ : PNP-N (n=20) ● : NOIGEN EA170

◑ : PNP-S (n=10) ◐ : NEOGEN R

⊗ : PNP-S (n=10) / PNP-N (n=20) = 3/7

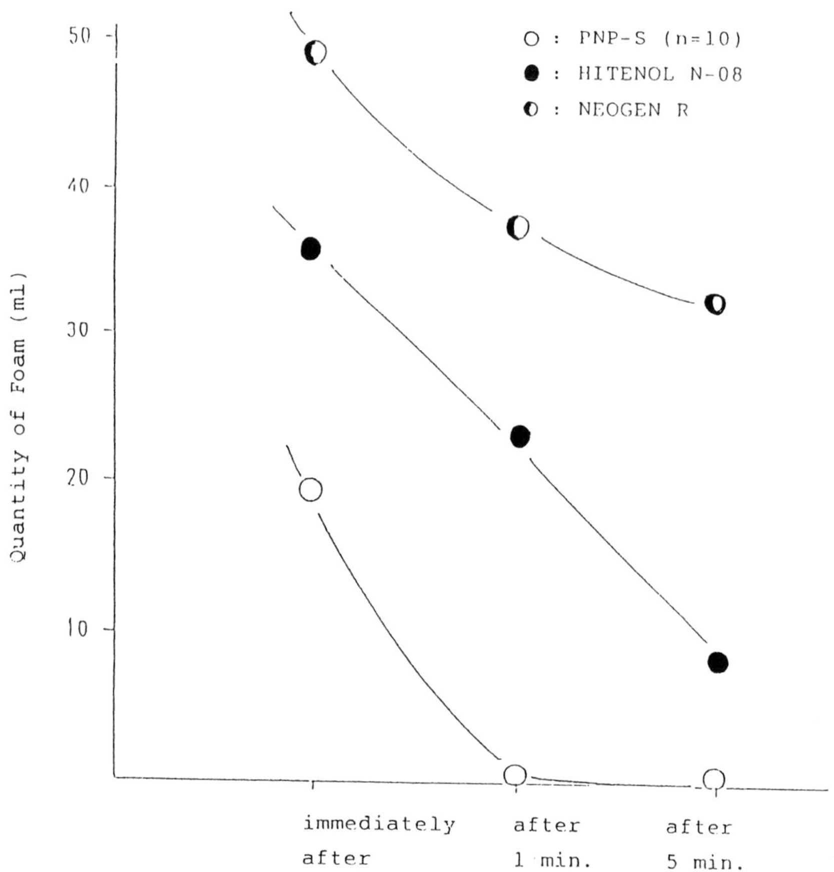

Fig. 5 Foaming of Emulsion (Styrene/Acrylic Acid Butyl)

(3) Water Resistance of Polymer Film

Water-resisting property was excellent in both polymer films of ethylacrylate and butylacrylate/styrene. Water resistance of any polymerizable surfactant is related to the degree of ethoxylation.

In the case of nonionic polymerizable surfactant, the greater the number of moles EO became, the more water-resisting property was lowered.
It was observed that both sulphate and phosphate types as anionic polymerizable surfactant had 10-20 moles of ethylene

oxide and this value is significant level. As the moles of ethylene oxide were decreased, water-resisting property was lowered. We believe that this is because sulphate and phosphate have stronger hydrophilic property than that of polyoxyethylene chain.

Table-11 Water-resisting Property of Emulsion

Polymer	Polymerizable Surfactant		Water Resistance (hr)
Ethylacrylate	AGNP-N	(n=20)	over 168
"	"	(n=50)	72
"	AGNP-S	(n=2)	16
"	"	(n=6)	120
"	"	(n=10)	over 168
"	"	(n=20)	over 168
"	PNP-P	(n=2)	24
"	"	(n=6)	over 168
"	"	(n=10)	over 168
Butylacrylate/ Styrene	AGNP-S	(n=2)	24
"	"	(n=6)	144
"	"	(n=10)	over 168
"	"	(n=20)	over 168

(4) Contact Angle of Polymer

When polymerizable surfactant was used for emulsion, high contact angle was observed in each polymer film of ethylacrylate and butylacrylate/styrene.

(5) Others

As for surface tension of emulsion in both ethylacrylate and butylacrylate/styrene, when anionic polymerizable surfactant was used, the value was approximately 6-8 dyne/cm and this is a significantly high value.

When polymerizable surfactant was used, mechanical stability was the same as conventional surfactant even though better stability had been expected to be observed.

3-5 Behavior of Polymerizable Surfactant

We believe that polymerizable surfactant in emulsion polymerization behaves as follows ;

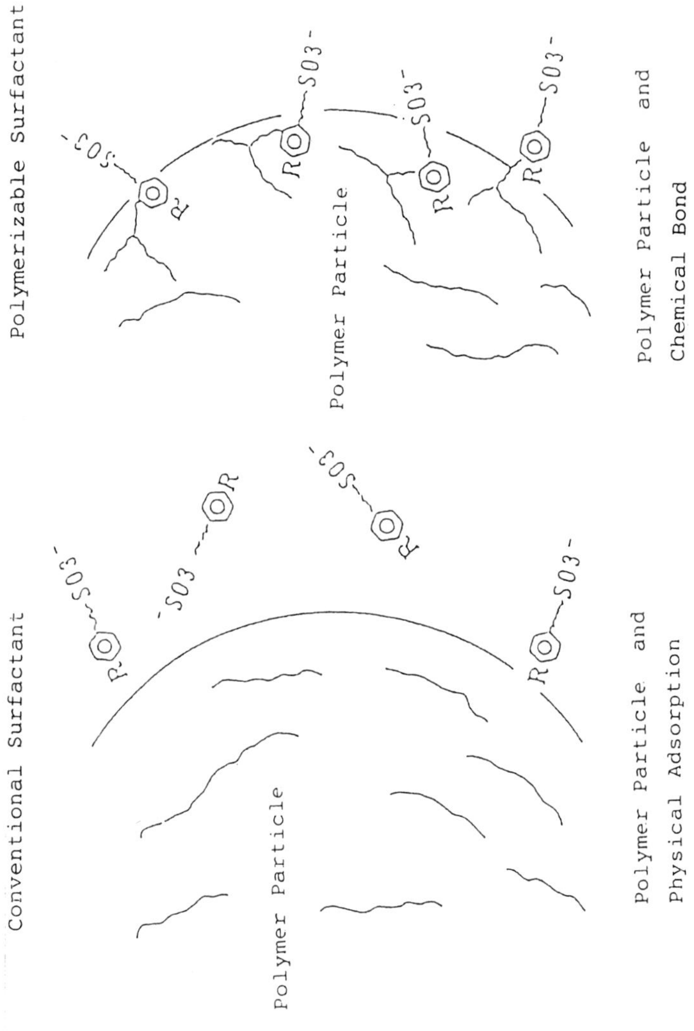

Fig. 6 Surface Conditions of Polymer Particle

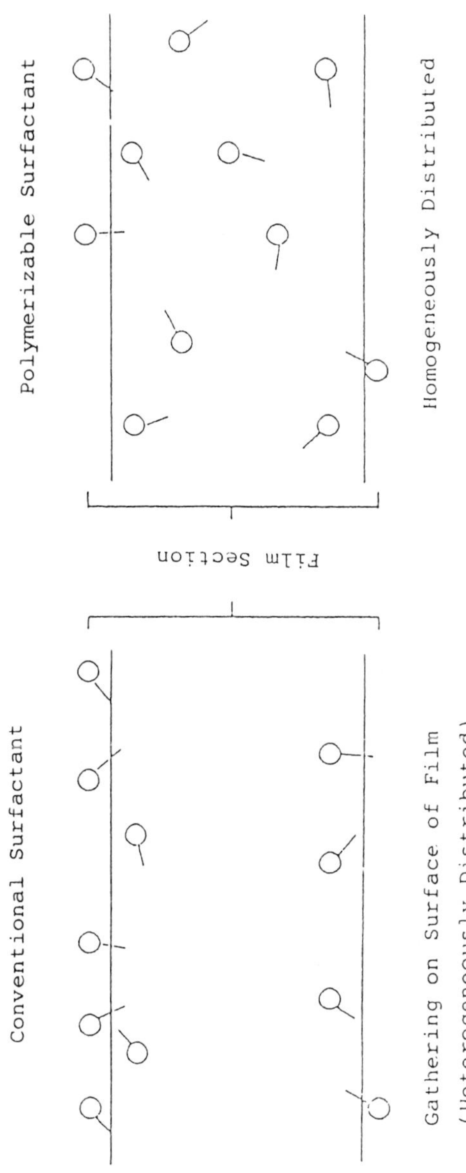

Fig. 7 Surfactant inside Polymer Film

Polymerizable surfactant forms micelles in water and provides an opportunity for polymerization. When monomer polymerizes, polymerizable surfactant also copolymerizes and it is taken inside polymer. In the early stage, a lot of copolymerized polymerizable surfactant exists on the surface of particle. The particle is kept stable, because polymerizable surfactant provided continuously copolymerizes and stays on the surface of particle. (Fig.6)

As a result, because an emulsion using polymerizable surfactant has little liberated surfactant, little foaming of emulsion and high surface tension are to be observed.

And as for polymer film, since polymerizable surfactant is distributed homogeneously and taken into polymer chain due to copolymerization, polymerizable surfactant is not distributed heterogeneously. (Fig.7) Therefore, better water-resisting property and higher contact angle are to be brought.

4. Conclusion

Polymerizable surfactant of nonion, sulphate-type anion, and phosphate-type anion were synthesized, and estimation was carried out as to polymerizable surfactant for emulsion polymerization.

Synthesis of polymerizable surfactant is sufficiently good to enable its production on a commercial basis, because the existing technology and facilities can be used and the yield of product is efficient.

As for the properties of polymerizable surfactant for emulsion polymerization, stability in polymerization was the same level as conventional surfactant. So it was proved that the biggest defects of emulsion, such as water resistance of polymer film and foaming of emulsion, could be improved.

References

1) Ohkubo, Matsumoto and others Journal of the Adhesion Society of Japan Vol.17 185 (1981)

2) Ohkubo, Matsumoto and others Journal of the Adhesion Society of Japan Vol.17 264 (1981)

3) Ohkubo, Matsumoto and others Japanese Journal of Polymer Science and Technology Vol.32 No.4 229 (1975)

4) Greene J.C.I.S. Vol.32 90, 96 (1970)

5) Yamazaki Surface Vol.25 No.2 86 (1987)

6) Sherrill.M.L., Matlack.E.S., J. AM. Chem. Soc., 59 2134 (1937

Comb Copolymers — A Unique Class of Polymer Surfactants

G. Krüger, A. van de Berg and E. Philippsen-Neu

AKZO — CHEMICALS DIVISION, RESEARCH CENTRE, 46 KREUZAUER STR. POSTFACH 100146, D-5160 DÜREN, GERMANY

1. INTRODUCTION

Due to often outstanding properties in the preparation/stabilization of disperse systems, polymer surfactants have met an increasing acceptance in a growing number of industrial applications during the last years.

Today already a number of product lines, each having specific advantages, are commercially available and well established for certain applications.

As it appears to be clear from theory, and there is also a lot of evidence from practice to support the assumption, that one single polymer will hardly meet the specific requirements in different systems in a satisfactory way, it is obvious that the molecular design and also the technology for the synthesis of polymer surfactants should be chosen carefully.
Highly desirable is a molecular architecture that allows a fair degree of flexibility with respect to the nature and the amount of anchoring and stabilizing groups. Among all structural types of copolymers that may show surfactancy, comb copolymers in this respect form a unique class of products.

2. STERIC STABILIZATION WITH COMB COPOLYMERS

The generally accepted mechanism for polymer surfactants having low if any charge density is that of "steric stabilization". Because of the fact that electrostatic stabilization normally fails in media with low dielectric constants due to a significant suppression of electric double layers, collodial stability in these systems, f.i. organic vehicles,may often only be achieved by steric stabilization.

For comb copolymers the situation at a liquid/solid interface may be outlined as in fig. 1.

A comb copolymer surfactant contains at least two classes of active sites: anchoring groups which are responsible for a sound attachment of the polymer to the surface of a solid colloidal particle, and pending sidechains which build up the barrier for steric stabilization. For effective stabilization of colloidal systems both functions, the anchoring as well as the stabilizing groups have to be specifically optimized.

Depending on the nature of the disperse phase, anchoring is normally provided by multiple noncovalent interactions. This can be Coulomb attractive forces of anionic or cationic groups that are complementary to charged groups on the surface of the dispersed particle, as well as hydrogen bonds or dipol-dipol interactions.

In certain cases, however, rather specific anchoring mechanisms are necessary to provide adsorption to the relatively inert surface of some organic colour pigments.

Though each of these bonds is relatively weak, the firm attachment is eventually achieved by the simultaneous action of a high number of these anchoring groups.

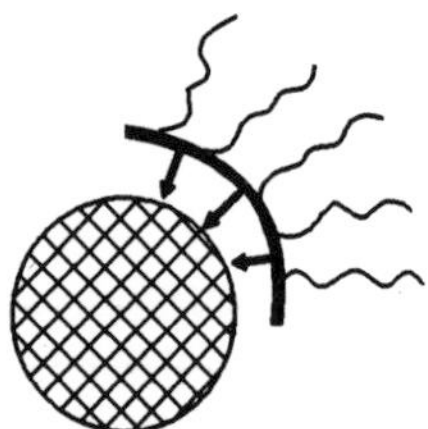

Fig. 1

Adsorption of a comb copolymer at a liquid/solid interface (schematically)

In the same way as the anchor groups should be complementary to the disperse phase, the stabilizing chains have to be adapted with regard to the continuous phase. For good steric stabilization, the sidechains should be readily dissolved by the continuous phase, in terms of polymer chemistry the continuous phase should be a "better than theta"-solvent for the stabilizing chains.

Though fig. 1 gives a good idea of the principle mode of action of comb copolymers, it is a rather optimistic description.
In reality, a polymer will hardly unfold in a way that all anchoring groups are directed towards the disperse phase and all stabilizing chains are completely solvated by the medium. As could recently be shown by experiments and by molecular dynamics calculations, comb copolymers may actually tend to form rigid rather unflexible conformations which allow only a very limited segment mobility under normal conditions. This limited segment mobility may of course significantly reduce the degree of unfolding and, as a consequence, may hinder effective site directed adsorption at the interface.

Macroscopic evidence for this correlation between structure and performance of comb copolymers may be observed experimentally. Fig. 2 is a representation of viscosity of a pigment slurry in an organic solvent at constant solids content.
The variable in these experiments was the relative segment mobility, hence the internal flexibility of the backbone of a specific type of comb copolymer. All other parameters, namely the nature of the anchoring and stabilizing groups and roughly also the overall polarity of the dispersant were kept constant.

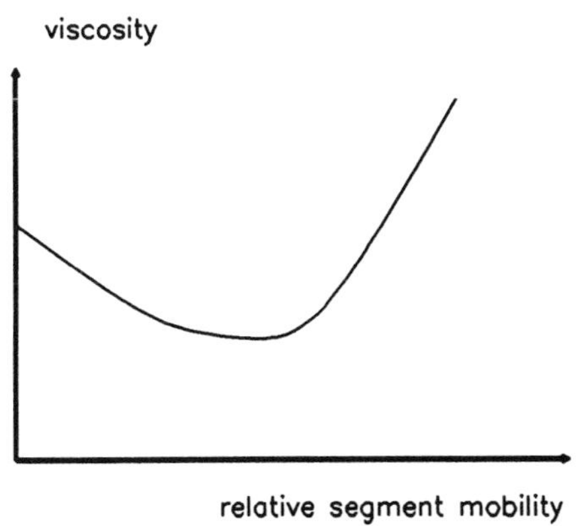

Fig. 2

Viscosity of a pigment slurry as a function of the polymer backbone.

Improving the segment mobility initially reduces the viscosity of the slurry; the function then reaches a minimum which represents the optimal degree of dispersion. Further enhancing the segment mobility leads to detrimental effects because further flexibilization goes along with a severe reduction of the steric stabilization barrier which is eventually reduced to below sufficient levels.

Thus the internal flexibility is the third parameter besides the nature of the anchoring and stabilizing groups that deserves attention and can be optimized easily in the case of comb copolymers, if the technology for the synthesis of these compounds is properly chosen.

3. DAPRAL^{R} - COMB COPOLYMERS

Akzo Chemicals has developed the Dapral^{R}-line of comb copolymers; the first generation of products, which is already commercially available is especially suitable to disperse/stabilize polar disperse phases in lower polarity vehicles (fig. 3).
This includes inorganic pigments, colour pigments and mineral fillers like silicates (talcum, mica) metaloxides, carbonates, hydroxides, carbon black and various others, but also liquid disperse phases like water or aqueous solutions.

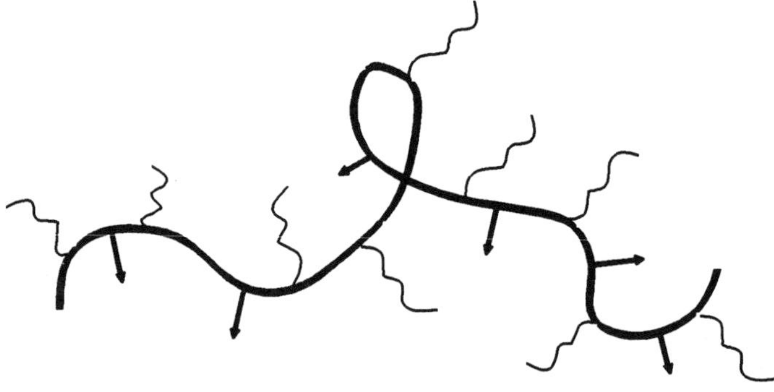

Fig. 3 Schematic representation of Dapral comb copolymers

In order to achieve the maximum effect with Dapral comb copolymers, the most suitable product should be carefully selected with regard to the solvent used (Tab. 1).
Whereas Dapral EN 1512 is most efficient in unpolar hydrocarbons, Dapral GE 202 covers the range of medium polarity solvents including alcohols and even water, EN 1511 is only recommended for use in highest polarity solvents.

Preselection of dispersant

	DAPRAL EN 1512	DAPRAL GE 202	DAPRAL EN 1511
Paraffins	//////		
Aromatics	//////	//////	
Esters	//////	//////	
Ketones		//////	//////
Alcohols		//////	//////

Tab. 1

Preselection for Dapral comb copolymers

The information given in this table may, however, only be seen as a preselection guide. In practice, these generalized recommendations can only help to select the most promising starting point; the dispersant that should be considered first in a given application. Thus it might be necessary to check also other candidates if the first choice is not satisfactory.

The advantages that may be observed upon use of Dapral comb copolymers may generally be related to a higher quality of the dispersion.

As a consequence, the rheological behaviour is improved which may facilitate processing or allow to increase filler load.

This effect is nicely documented upon use of comb copolymers as dispersants in resin-free millbases of colour pigments (Fig. 4). Whereas the viscosity of a conventional millbase containing 20 % of a long oil alhyd resin sharply increases at loads of Red 179 higher than 40 %, the millbase containing the comb copolymer without additional resin is still fluid up to concentrations well above 50 %.

The higher quality of dispersion is eventually also reflected in the appearance of endproducts; smoothness and gloss of surfaces for instance is generally improved.

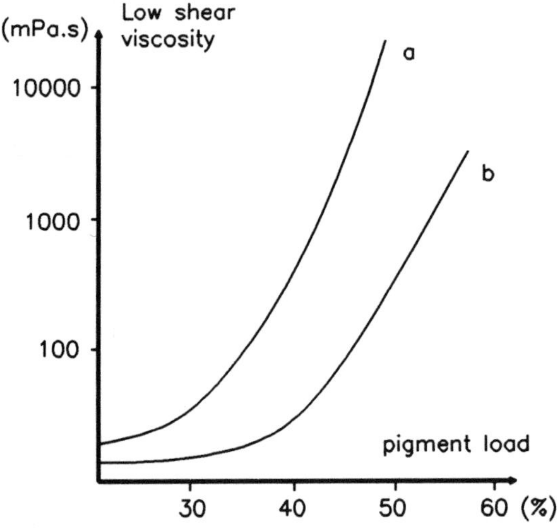

Fig. 4

Low shear viscosity of P.R.179/ white spirit mill base as a function of pigment load.

a: conventional long oil alkyd resin mill base

b: resin free mill base containing a comb copolymer dispersant.

4. APPLICATIONS

From the large number of possible applications for Daprals that already became apparent, three specific examples are discussed in more detail in the following.

4.1. Processing aids for mineral filled polypropylene

A particular representative of the Dapral^{R} range is marketed in the plastics compounding and processing industry under the tradename ARMOWAX W 440. This specific type of comb copolymer proved to be a highly efficient wetting/dispersing agent for all kinds of fillers in polyolefins, and other thermoplastic matrices.

Taking for example the extrusion of polypropylene with a 40 % load of talcum, the degree of particle agglomeration is remarkably reduced upon addition of Armowax W 440 (Tab. 2). This is interalia well documented by electron microscopy investigations of the final compound (fig. 5).

This is a REM picture of a 40 % talcum filled polypropylene, compounded on a laboratory twin screw extruder.

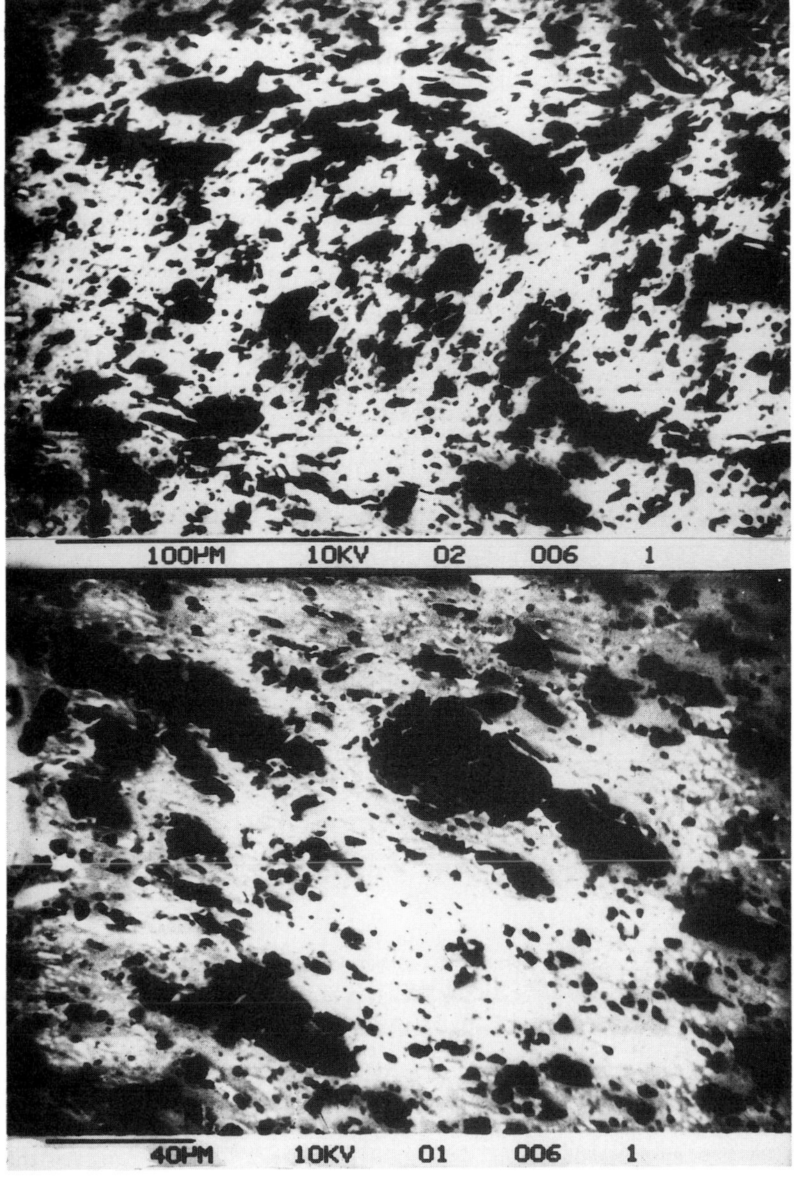

Fig. 5 REM-picture of a polypropylene compound (Propathene GY 621 M, ICI) containing 40 % talcum (grade 1445, Luzenac) Top: blank
Below: 2 % ARMOWAX W 440

The blank obviously still contains larger agglomerates of the filler due to insufficient wetting and dispersion. Adding 2 % W 440 clearly improves the situation (due to an effective wetting and dispersing) nearly all agglomerates are broken.
As a consequence of the improved deagglomeration of the filler the melt flow index rises from 1.6 to 2.5 g/min. The higher quality of dispersions of the filler is also reflected in the properties of the final compound. Thus elongation at break is nearly doubled upon addition of Dapral, whereas the reduction of tensile strength is negligible, showing that Armowax W 440 is hardly influencing the softness of the compound and does not act as a plasticizer.

Property	Blank	2% ARMOWAX W440
Content of agglomerates		reduced
Elongation at break	6.7 %	11.5 %
Tensile strength	20.7 N/mm ^{2}	18.8 N/mm ^{2}
Impact strength (ISO 180)	21 kJ/mm ^{2}	23.1 kJ/mm ^{2}
Surface gloss (20°)	16.4	23.3
Melt flow index	1.57 g/min	2.45 g/min

Tab. 2 Processing- and mechanical properties of a polypropylene compound (Propathene GY 621 M, ICI) containing 40 % talcum (grade 1445, Luzenac)

Eventually the appearance of the whole compound, surface smoothness and gloss of injection molded parts f.i. is very positive. These advantages become even more obvious if the compounds are processed on large industrial scale extruders that may run at a much higher processing rate than the laboratory twinscrew extruder that was used for this investigation.

4.2. Dispersants for paints/coatings/ink applications

Coatings is another area where polymer surfactants may be used advantageously. Dapral GE 202 is highly suitable for dispersion of fumed silica, which is frequently used as a matting agent in all kind of resins and varnishes.

In an acid cured amino alkyd coating (Tab. 3) colloidal stability of the system is achieved by addition of 0.5 % Dapral GE 202. Whereas the initial viscosity of the formulation is hardly influenced (silica content is 8 %) the higher quality of the dispersion becomes evident upon storage.
The blank after one month shows a hard, poorly redispersible sediment, whereas the Dapral GE 202 containing formulation shows no sedimentation at all.

The clear supernatant makes 85 % of the total volume in case of the reference and only 25 % if 0.5 % Dapral GE 202 are added.
Eventually, as a result of the high degree of deagglomeration, the matting efficiency of the silica is improved by nearly 50 %.

Property	Blank	0.5% DAPRAL GE 202
Viscosity Brookfield 6rpm initial (mPa.s/cP)	300	340
Stability after 1 month	hard sediment	no sediment
clear supernatant after 1 month	85 %	25 %
Film gloss, 60° (matting efficiency)	12	7

Tab.3 Colloidal stability of an acid curable aminoalkyd resin containing 8 % b.w. of an untreated 4.5 μm silica as a matting agent.

It is important to mention that the curing reaction, acid cure in the case of an aminoalkyd-coating or UV cure in the case of acrylate resins, is unaffected by Dapral GE 202.

4.3. Inverse suspension/emulsion polymerization

As a last example I would like to discuss some preliminary results that have been obtained by using comb copolymers as an emulsifier/stabilizer system in inverse suspension polymerization of sodiumacrylate. The end product of the ISP-process is a highly swellable polyelectrolyte gel which found major use in personal care products as for example baby diapers.

The general advantage of the inverse suspension technique is that ready-to-use polymer beads are formed without a final grinding step that often produces a high share of unfavourable fine dusty particles. In order to ensure colloidal stability of the inverse suspension during polymerization, a low HLB emulsifier like sorbitanemonooleate is normally added in amounts of up to 2 %.

Upon use of polymer surfactants the key properties of the superabsorbing polymer beads may be significantly influenced as could be shown recently by Trijasson and coworkers[1)2)] at the Polymer Institute in Strasbourg, who investigated A-B-A-block type copolymers.
First results that we obtained in our laboratories with Dapral-type comb copolymers confirmed the idea that dedicated polymer surfactants may offer new opportunities in this area.
Depending on the nature and the concentration of the comb copolymer emulsifier uniform spherical particles with a diameter in the range of 20 - 200 μm could be synthesized, whereas with conventional low molecular weight emulsifiers like sorbitanemonooleate particle size is restricted to below 50 μm.

Particle size and uniformity are important issues. Large uniform particles are highly desirable, because they allow easy and safe handling and normally exhibit a high moisture retention in the swollen state even under pressure which of course is of major importance for baby diapers. On the other hand extremely large particles suffer from poor uptake kinetics and low equilibrium levels, which is frequently also deteriorated by the presence of the hydrophobic low HLB emuslifiers that are used.
So, commonly a compromise is made which takes into account both aspects, handling and performance.

The next picture (fig. 6) shows beads with an average size of about 80 μm and a fairly narrow distribution that have been synthesized using Dapral comb copolymers.

Though our investigations have not yet been brought to an end, it is obvious that dedicated comb copolymers offer a wide room for specific adjustment of properties of products derived from inverse suspension and emulsion polymerization.

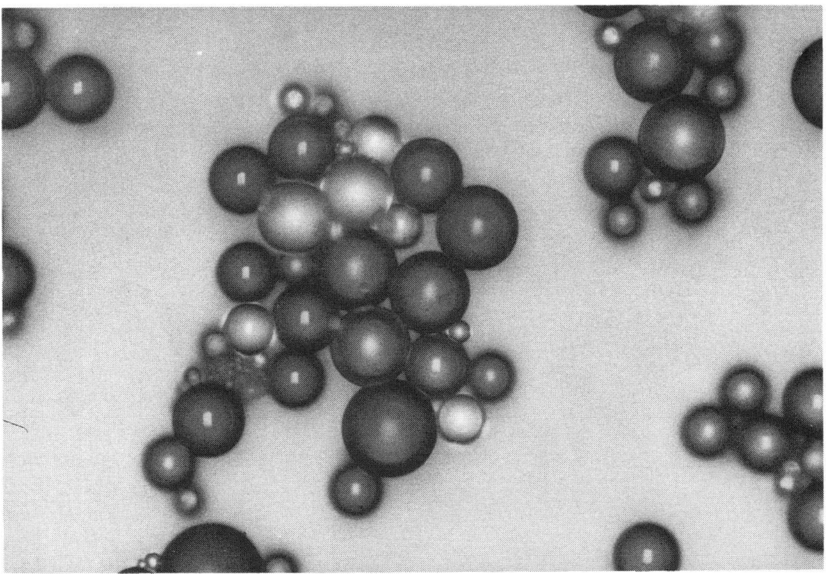

Fig. 6 Optical micrograph of sodiumpolyacrylate-gel beads, obtained by inverse suspension polymerization using a comb copolymer emulsifier.

Aqueous phase	:	40 % sodiumacrylate, 0.26 % methylene bisacrylamide
Organic phase	:	cyclohexane: w/o = 1/2 (v/v)
Initiator	:	$K_2S_2O_8$
Average particle size	:	80 μm

5. CONCLUSION

The current range of Dapral comb copolymers are effective polymer surface active agents which may be used advantageously to disperse polar particles/droplets in various matrices. Advantages are strongly dependent on the relevant applications; generally the rheology of the system is improved, which allows to increase the load of disperse phase and/or facilitates processing.
Furthermore an increased colloidal stability upon storage or under demanding conditions can be expected. Endproduct properties, as a consequence of a high quality dispersion, may be drastically influenced.

Possible applications are in processing of all kinds of thermoplastics (filled/pigmented/-reinforced systems) in ceramics processing in coatings, inks sealants, w/o-emulsions, inverse polymerization processes and other applications.

Our development activities are now concerned with a second generation of comb copolymers which are more dedicated to organic colour pigment dispersions. First results encouraged us to continue research on comb copolymers.

6. REFERENCES

1) P. Trijasson, T. Prith, M. Lambla, Makromol. Chem., Macromol. Symp., 1990, 35/36, 141.

2) P. Trijasson, Y. Frère, P. Gramain, Makromol. Chem., Rapid Commun., 1990, 11, 239.

Anionic Surfactants

The Structure, Performance and Environmental Aspects of Linear Alkylbenzene Sulphonate

R. Fellows and F.W. Heywood

SHELL CHEMICALS U.K. LTD, STANLOW MANUFACTURING COMPLEX, ELLESMERE PORT, SOUTH WIRRAL, L65 4HB, UK

Introduction

As the RSC was looking forward to it's centenary celebrations the world was at war. The developing petrochemical industry was focused on maximising the production of fuels, particularly aviation fuel. Upgrading propene by production of the trimer and tetramer to produce olefins and paraffins with high branching and therefrom high octane components was priority work. Following the war other outlets for the product were sought. The sulphonation of alkylate produced a synthetic detergent with excellent properties. The highly branched chains gave excellent chemical stability for surviving the rigours of the manufacturing cycle but also resulted in the first foaming incidents in waste water treatment plants as early as 1947.

It was quickly learned from the earliest studies on biodegradation that the structure which had such good cleaning properties, could not be degraded rapidly enough in waste water treatment plants. Study groups examining the many scientific and technical issues discovered that straight chain carbon polymers, much like the fatty acid chains of soap, could be readily biodegraded. This provided the clue to answering the environmental challenge of the day.

Following the introduction of LAS into commerce, a significant reduction in surfactant levels in waterways and in foaming incidents was readily apparent. Since then LAS has been the predominant surfactant in commercial detergent preparations. Production figures for linear alkylbenzene worldwide are shown in Table 1. Yearly production in USA, Japan and W. Europe is approximately 1.7 million tons (over 2 million tonnes as LAS). It is the single largest, and arguably the best, synthetic detergent base in use today.

Table 1 Production Figures for Linear Alkylbenzene Worldwide 1989 (k.tonnes)

Western Europe	417
Eastern Europe	214
N. America	320
Asia	353
South Asia	130
Africa/Middle East	153
Americas (excl USA & Canada)	134
Total	1721

LAS is probably the most intensively studied molecule in production. Its biodegradational behaviour has been comprehensively studied. It is basically an environmentally compatible product whose ecological effect, with todays usage, is identified and controllable. The concentration treated in sewage works has been estimated to be 35% lower than the quantity expected from total use and population. The difference can probably partially be explained by the loss of LAS due to biodegradation en route to the treatment system.

The removal of LAS in efficiently operated sewage treatment works is high (95-99%). The proportion of LAS associated with sewage sludge is approximately 30% as LAS is not biodegraded under strict anaerobic conditions. However, any LAS present in sludge-amended soils degrades at relatively high rates with half lifes ranging from 3 to 35 days.

Evidence from many studies[(1)] shows that under aerobic conditions LAS is completely degraded (mineralised) in the environment. By any reasonable standard the degradation of LAS under aerobic conditions is proven.

The Chemistry of LAB and LAS

LAB is a family of related compounds that results from the chemical synthetic processes used to manufacture it. The variations on the theme are (1) the lengths of the carbon chain in the final product, and (2) the position in the carbon chain where the benzene ring ultimately resides (Figure 1).

Table 2 shows the carbon chain distribution of two typical manufactured LAB mixtures. One has predominantly shorter carbon chain components (C9 and C10), while the other has mainly longer carbon chain (C12 and C13). Nonetheless, most LAS preparations have mixtures of linear carbon chains of lengths between 10 and 14 carbons. The consequences of the presence of mostly longer or shorter carbon chains are discussed later.

Table 2 Typical Carbon Distributions of Two Commercial Linear Alkylbenenes

Carbon Distribution % m/m		C_8-C_{13}	C_{10}-C_{13}
Phenyl-	C_8	1	
Phenyl-	C_9	16	
Phenyl-	C_{10}	27	10
Phenyl-	C_{11}	22	35
Phenyl-	C_{12}	21	35
Phenyl-	C_{13}	14	20
Mean Mol Mass		230	241

The major peaks on the chromatogram of LAB are the phenyl isomers. Note that there is no 1-phenyl isomer present. The minor peaks in the chromatogram have been identified by GC/MS as tetralins and branched alkylbenzenes[(2)]. The nature and level of these entities are dependent on the manufacturing route used, and cannot be separated by distillation. We need therefore to consider the manufacturing processes involved in the production of LAB.

Figure 1

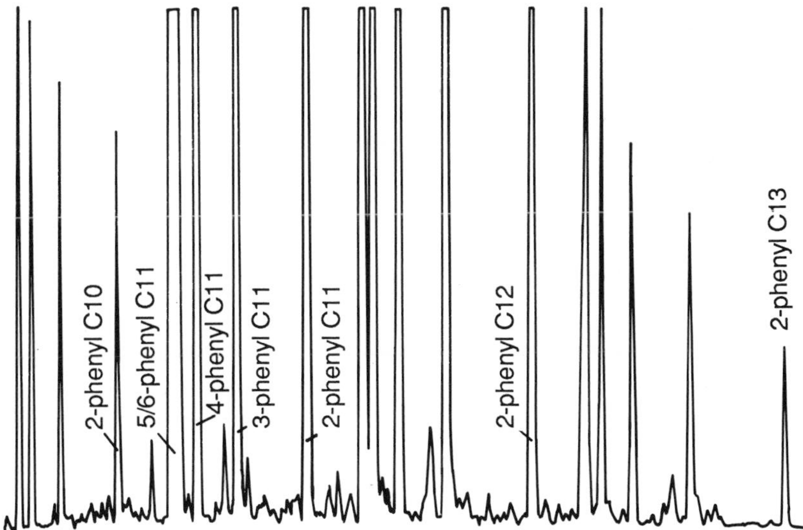

The Manufacture of Linear Alkylbenzene

Detergent alkylate (linear alkylbenzene) is manufactured by a Friedel-Crafts catalysed alkylation of benzene with either aluminium chloride or HF. Linear alkenes, obtained either by dehydrogenation of n.alkanes, extracted from kerosene, or derived from ethene by proprietary processes or obtained indirectly by chlorination/dehydrochlorination of n.alkanes are used as feedstock. Two commercial processes are descibed schematically in Figure 2. Almost two-thirds of LAB is from the hydrogen fluoride process. Most of the alkylate used in liquid products is probably from the aluminium chloride process; the situation is just the reverse in powders.

The alkylbenzenes obtained are not the same. They differ in isomer distribution with the phenyl group being distributed differently along the chain. The greatest difference is in amount of 2-phenyl isomer formed as shown for a hypothetical dodecylbenzene:

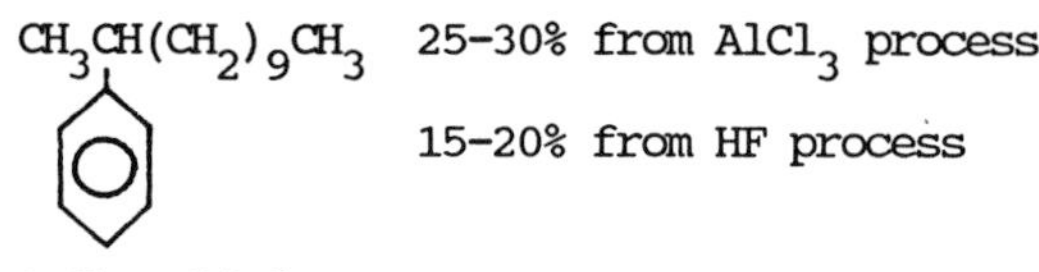

2-Phenyldodecane

The $AlCl_3$ catalysed process gives rise to about 25-30% of the 2-phenyl isomer with the amounts of 3,4,5 and 6-phenyl isomer decreasing steadily from about 22% to 15% respectively. The HF process gives an LAB with about 15-20% of the 2-phenyl isomer with the others distributed fairly evenly among the secondary carbon atoms.

The alkylates obtained from the $AlCl_3$ and HF processes are, therefore, commonly referred to as high and low 2-phenyl LAB, respectively, as are the LAS's (Table 3). These differences in composition markedly effect formulating properties of LAS in liquid detergents, particularly solubility and viscosity.

Chemistry for Alkylation of Benzene

$$\text{(i)}\quad \text{Olefin} + \text{Benzene} \xrightarrow{\text{HF}} \text{Alkylbenzene}$$

$$\text{(ii)}\quad \text{Monochloroparaffin} + \text{Benzene} \xrightarrow{AlCl_3} \text{Alkylbenzene} + \text{Hydrochloric Acid}$$

Figure 2

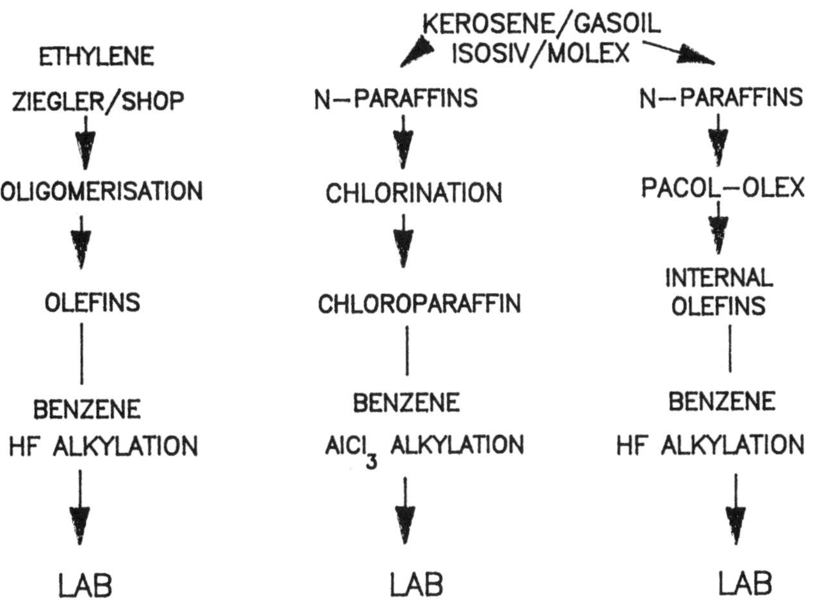

A very low tetralin content indicates that LAB was produced via HF process from alkenes low in dienes. Dienes in cracked wax alkenes and dichlorinated compounds in chloroalkanes are believed to be the source of alkyltetralins and alkylindanes in linear alkylbenzene. Dienes can also give rise to diphenylalkanes and heavy ends. It is therefore to be expected that the use of high quality alpha or internal olefins would give a linear alkylbenzene containing lower levels of these byproducts.

Table 3 Analysis of Alkylate from HF and $AlCl_3$ Processes (GC-MS)

		HF	$AlCl_3$
2 Phenyl Alkanes	% m/m	17.0	28.8
Tetralins	"	0.1	11.7
Branched alkylates	"	2.5	1.2
Diphenylalkanes	"	0.3	-
Molecular weight		241.0	244.0

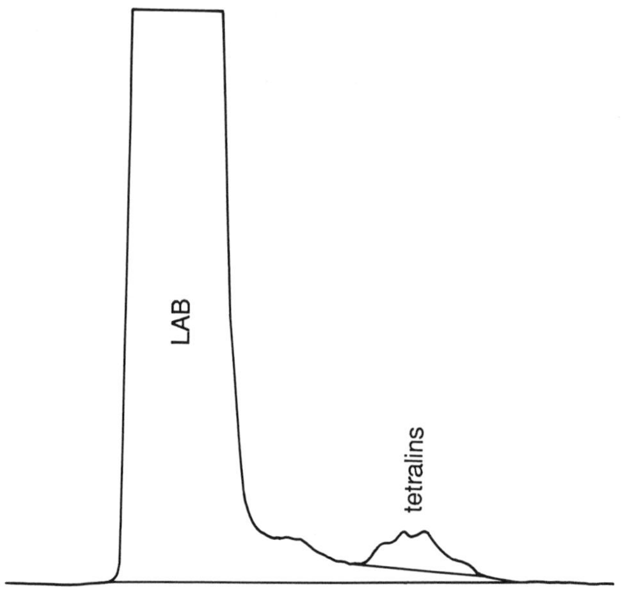

Figure 3

HPLC Chromatogram of Typical Commercial Linear Alkylbenzene

An analytical method, based on an HPLC procedure proposed by Cavalli[3], for the determination of dialkyl tetralins is currently being reviewed by an ECOSOL Analytical Task Force (Figure 3).

Sulphonation Chemistry

Treatment of an organic compound containing an aromatic nucleus with a strong sulphonating agent results in the substitution of one of the nuclear hydrogen atoms by a sulphonic acid group. In the case of alkylbenzenes reaction takes place in the ortho and para positions, mainly the latter.

C_nH_{2n+1} (benzene ring) $+SO_3$ (or $+H_2SO_4$) -------> C_nH_{2n+1} (benzene ring) SO_3H $(+H_2O)$

Because of side-reactions leading to by-products such as sulphones etc, it is never possible to achieve 100% conversion of alkylbenzene into sulphonic acid. Under controlled conditions, however, yields in excess of 98% can be obtained (as assessed by the amount of residual unsulphonated material).

Analysis

The accepted technique for accurately determining the purity of a nominally pure compound is to assay the impurities. This is the concept behind the "difference" method for determining AM content of alkylbenzene sulphonic acids.

The three known impurities (free oil, H_2SO_4 and H_2O) are all present in sufficiently low concentration to enable the results obtained, within the accuracy of the methods used, to still have better reproducibility for AM content than the results of direct methods (eg. ISO 2271).

This method of determination assumes that only three major impurities are present. Direct titration methods have the following disadvantages:

1. They rely on accurate equivalent weights which can only be obtained from exact known data of the parent alkylate.
2. The published reproducibility data are greater than variations in AM content which are acceptable to many customers.
3. They tend to give lower results than the "difference" method.

Sulphonation in Practice

The reagents which are normally used are as follows:-

(a) Sulphur Trioxide
(b) Oleum (various strengths)
(c) 98% Sulphuric Acid

The sulphonation can be carried out as a batch or a continuous process. The main variables which are likely to have a significant effect on the quality of the product are:-

1. The ratio sulphonating agent/alkylate
2. Sulphonation temperature
3. Reaction time (a combination of addition time and digestion time for a batch process)
4. Time and temperature in the acid separation stage (not for SO_3 technique)

As can be seen from the chemical equation water is produced when sulphuric acid is employed as the sulphonating agent. This water has a diluting action and may eventually stop the reaction before it reaches completion.

To overcome this problem, it is necessary, when using sulphuric acid or oleum as the sulphonating agents, to employ them in considerable excess over the theoretical quantity. Thus, increasing the severity of the sulphonation conditions is desirable to maximise the yield of sulphonate, but will tend to affect colour and possibly also odour adversely.

To achieve the lowest inorganic sulphate content requires a minimum of free sulphuric acid in the product before the latter is neutralised with alkali. This particular product requirement is not so important if the sulphonate is intended for use in detergent powders in which sodium sulphate can form part of the formulation. In light duty liquids and laundry liquids, however, it is very important because sodium sulphate has an adverse effect on solubility and low temperature stability.

Table 4 gives a summary of the suggested conditions for the sulphonation of LAB with sulphur trioxide, oleum and sulphuric acid.

Acid Ageing

Alkylbenzene sulphonic acids are slightly unstable during storage. This is evidenced by increasing colour, decreasing unreacted organic matter and free sulphuric acid.

The majority of this post reaction is complete after 24 hours but some reaction continues for 14 days. The by-products of the sulphonation reaction are pyrosulphonic acids and sulphonic anhydrides. At reactor temperatures below 35°C and with excess alkylate, these decompose to form sulphonic acid and yield a colour stable product.

Table 4 Suggested Sulphonation Conditions for Linear Alkylbenzene

Sulphonating Reagent	Quantity of Reagent Required Molar Ratio	Addition Temp °C	Digestion Temp °C	Digestion Time mins.
Gaseous SO_3	1.02/1	50-55	55	5-20
22.5% oleum	2.9/1	50-60	40	90
98% H_2SO_4	4/1	45-55	55	180-210

Figure 4

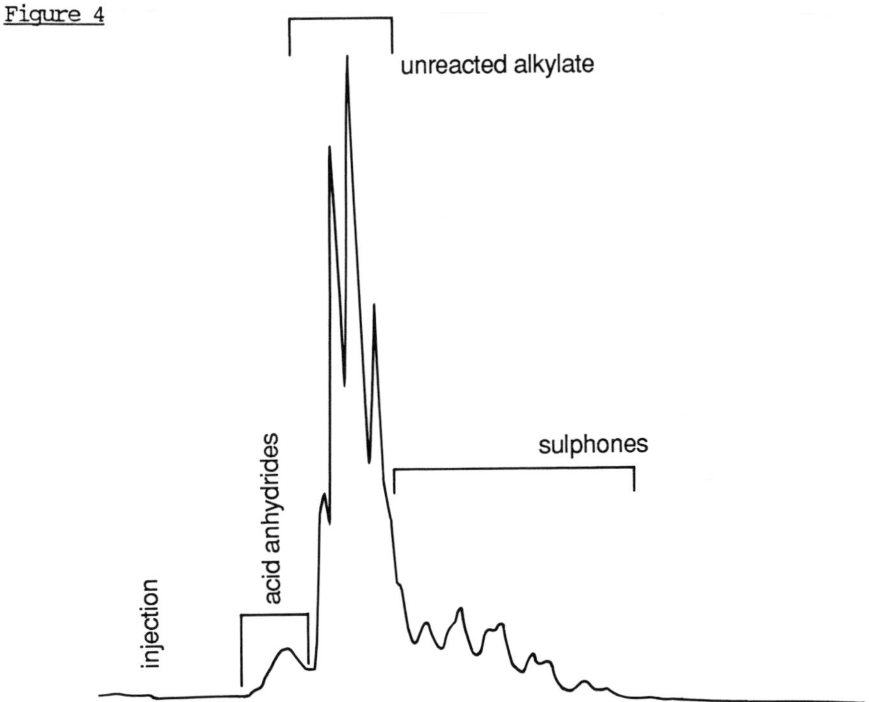

HPLC Chromatogram of Free Oil in Unaged Alkylbenzene Sulphonic Acid

Using a combination of mass spectrometry and SO_2 fluorometry, Petresa(4) identified another set of by-products in the sulphonation product, sulphones. They postulated a mechanism for their formation by reaction of pyrosulphonic acid with alkylbenzene. They found that a high sulphone content had a negative effect on the solubility characteristics of sodium sulphonate solutions.

In order to detect these components the pentane extracted fractions were dissolved in methanol and analysed by reversed phase HPLC using UV detection at 220nm. The results obtained (Figure 4) showed that the extracts contained groups of compounds not found in the parent alkylate.

Analysis of the product from Shell Chemicals operated plants has shown that when water had been added the concentration of components detected was markedly reduced. Sulphonic anhydrides, present in the acid reaction mixture, is reacted with water to form sulphonic acid. The acid anhydrides can be identified by the HPLC procedure.

The group of peaks eluting after the unreacted alkylate are sulphones There should be 7 sulphone peaks in all for a C10 to C13 alkylate. Due to the possibility that some peaks of the same carbon number are formed from different combinations of alkyl groups these peaks are doublets.

<u>The Reaction Mechanism</u>. One source of sulphuric acid present in the acid reaction mixture is derived from traces of water in the alkylate and dilution air.

$$SO_3 + H_2O \rightarrow H_2SO_4 \qquad (1)$$

One proton from sulphuric acid is more readily dissociated in the reaction mixture. This is also evidenced when titrating sulphonic acid in propan-2-ol with cyclohexylamine.

$$H_2SO_4 \rightarrow H^+ + HSO_4^- \qquad (2)$$

Petresa proposed that the site of acid attack on pyrosulphonic acid would give rise to sulphone formation or sulphonic anhydride.

$$\underset{\text{pyrosulphonic acid}}{R(SO_2)O(SO_3H)} + \underset{\text{alkylate}}{RH} \xrightarrow{H^+} \underset{\text{sulphone}}{R(SO_2)R} + H_2SO_4 + H^+ \qquad (3)$$

$$\underset{\text{pyrosulphonic acid}}{R(SO_2)O(SO_3H)} + \underset{\text{alkylate}}{RH} \xrightarrow{H^+} \underset{\text{anhydride}}{R(SO_2)O(SO_2)R} + H_2O + H^+ \qquad (4)$$

The water generated by anhydride formation can react with entrained SO_3 as shown in equation (1) or with the anhydride to form sulphonic acid.

$$\underset{\text{anhydride}}{R(SO_2)O(SO_2)R} + H_2O \rightarrow 2\ \underset{\text{sulphonic acid}}{R(SO_3H)} \qquad (5)$$

The anhydride can also react with sulphuric acid to form sulphonic acid and pyrosulphonic acid.

$$\underset{\text{anhydride}}{R(SO_2O(SO_2)R} + H_2SO_4 \rightarrow \underset{\text{sulphonic acid}}{R(SO_3H)} + \underset{\text{pyrosulphonic acid}}{R(SO_2)O(SO_3H)} \qquad (6)$$

A proposed mechanism for sulphone and sulphonic anhydride formation from pyrosulphonic acid, based on the acidity arguments put forward by Petresa, is given in Figure 5.

Figure 5

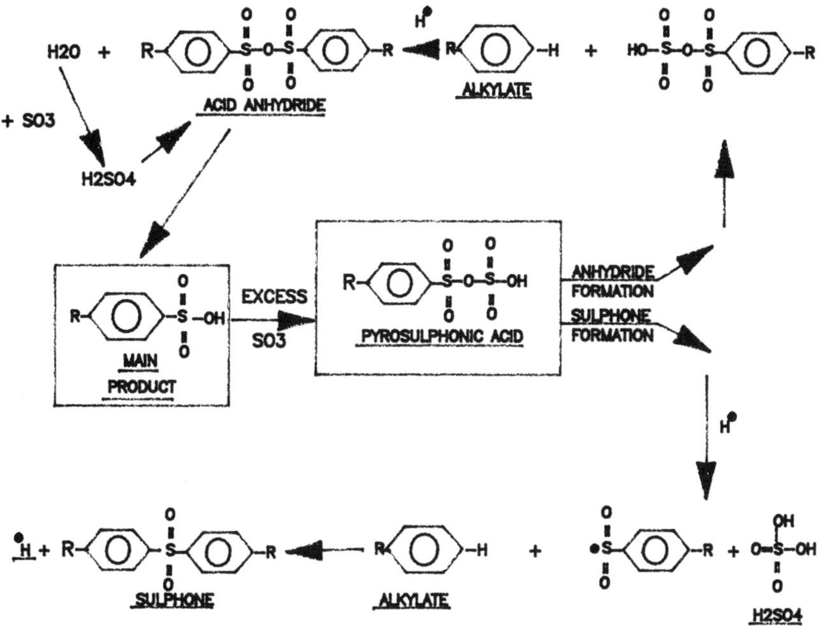

When samples of unaged acid were analysed, the ratio of anhydrides to sulphones was approximately 2:1, assuming equal molar extinction coefficients. This indicates that the anhydride formation mechanism predominates.

Petresa had reported that sulphone formation was dependent on the SO_3:alkylate ratio. The mechanism in Figure 5 is consistent with this assertion. More SO_3 will form more pyrosulphonic acid which will form more sulphone. More sulphonic anhydride is also formed but since this inevitably becomes sulphonic acid in an aged sample, the sulphone content is dependent on SO_3:alkylate ratio.

Physical Properties of Linear Alkyl Benzene Sulphonate

Effect of Minor Impurities in Feed and Product

Free oil content in linear alkylbenzene sulphonates has a considerable influence on the physical properties (viscosity and cloud point)[5]. It has been found that a determining factor in this influence is the sulphone content in free oil.

Low tetralin content alkylates, on sulphonation, show increasing viscosity with increasing SO_3/LAB ratio. Tetralins depress the effect because of the general viscosity cutting effect of tetralin sulphonates.

Neutralisation

Whichever reagent is used for sulphonation, it is necessary to neutralise the sulphonic acid before it can be incorporated into a liquid or powder detergent. Sodium sulphonates are by far the most commonly used although ammonium, calcium, magnesium, monoethanolamine and triethanolamine sulphonates are also used in certain speciality products. The calculated quantity of neutralising agent is prepared as a solution and charged to the neutralisation vessel. The sulphonic acid is added as quickly as possible, while cooling and stirring, without allowing the temperature to exceed 55°C. Neutralisation may take up to four hours, depending on the amount of residual sulphuric acid.

The alkyl chain length, the neutralising agent and the presence of tetralins have an important influence on solubility, viscosity and surface tension. When comparing the properties of the sodium and alkanolamine sulphonates, however, it must be remembered that the equivalent weights are significantly different; thus, a 30% active matter solution of the sodium sulphonate and a 41% active matter solution of the triethanolamine sulphonate derived from a sulphonic acid based on linear alkylbenzene with C8-C13 alkyl chains are about equivalent in anionic sulphonate content.

Effect of Neutralising Counterion

The effects of different inorganic and organic counterions on the physiochemical behaviour of three commercial linear alkylbenzene sulphonates have been studied[6]. It has been found that the counterion hydration radius has great influence on solubility, viscosity, surface tension and critical micelle concentration (CMC). The larger the ionic radius of the counterion, the lower the solubility regardless of the starting LAB used (NH_4 <K <Na <Li solubility increase). As expected, the increase in solubility has a direct relationship with the polarizing strength of the counterion and therefore with its hydration ease.

Viscosity/Temperature Relationship

The viscosity of linear alkylbenzene sulphonic acid changes markedly with temperature, as shown in Figure 6. Below about 20°C it is not very mobile, but on warming to 40°C, the viscosity is greatly reduced.

Figure 6

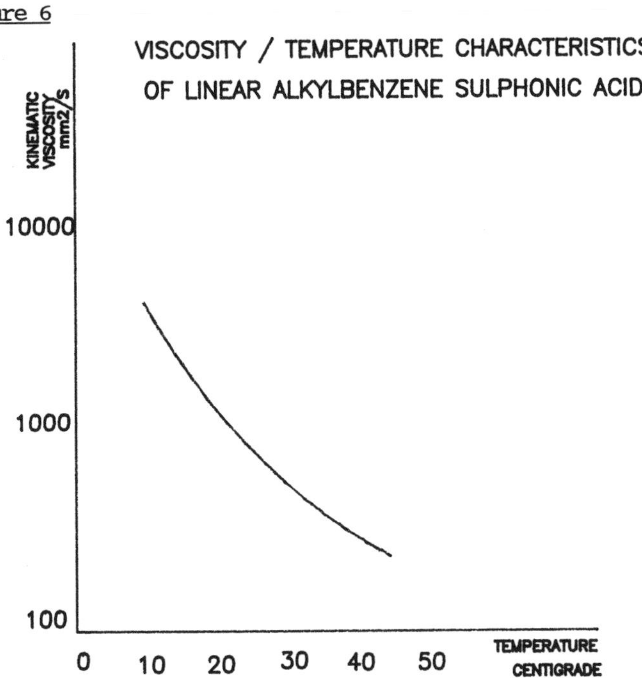

Viscosity of the sulphonate follows the opposite direction of solubility. The larger the counterion size, the more viscous the sulphonate solution: Li <Na <K <NH_4.

Tetralins reduce viscosity particularly for the organic salts where the more ethanol groups in the molecule the lower the viscosity.

Performance Aspects

Shell experience with dishwashing products indicates that foam stability and foam propensity are the most important performance aspects. Sulphonic acids with an average molecular weight corresponding to a C11-C12 side-chain is particularly recommended as a component of dishwashing and general cleaning liquids where its high solubility enables excellent cold stability characteristics to be obtained (Figure 7).

Interactions with water hardness ions decrease detergency performance. The detrimental effects are widely known to occur with LAS by the precipitation of it with calcium ions.

Figure 7

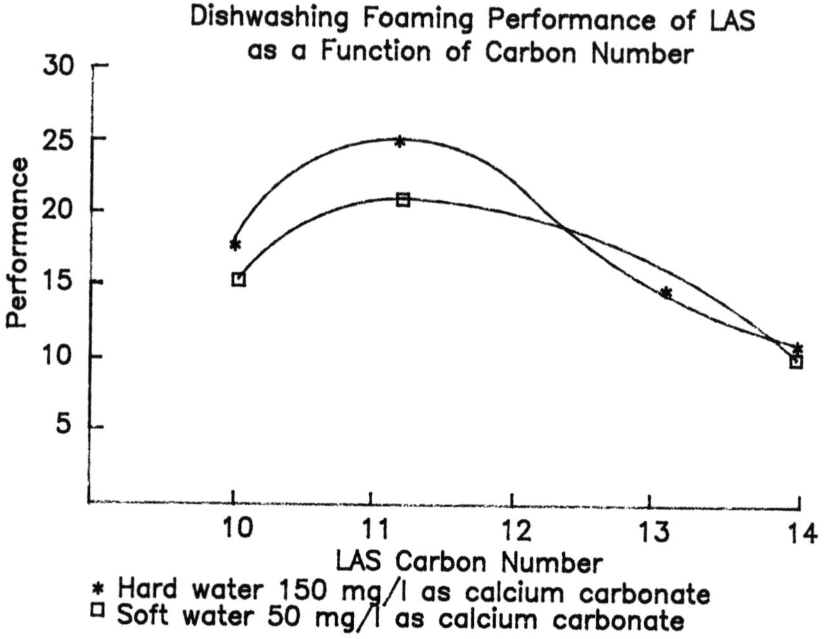

It is generally accepted that maximum fabric detergency requires a somewhat higher molecular mass product. Studies in Shell laboratories have shown that the optimum chain length for both cotton and synthetic fabrics is somewhat longer but is limited by Ecotox considerations to LAB rich in the C11-C13 carbon range (Figure 8).

In order to examine the interactions of LAS with calcium ions over a range of both surfactant and hardness concentrations, precipitation boundary diagrams are constructed. These diagrams provide much useful information about the interaction of anionic surfactants with Ca^{2+} ions[7].

A generalised diagram of the boundary is shown in Figure 9. Theoretically, the line segment SPL, or surfactant precipitation line, represents the solubility limit of surfactant monomers in the presence of Ca^{2+}.

The curved line segment MSL is the micelle saturation line. Above this boundary, as surfactant concentration increases, precipitated calcium surfactant is resolubilised by the micelles.

Figure 8

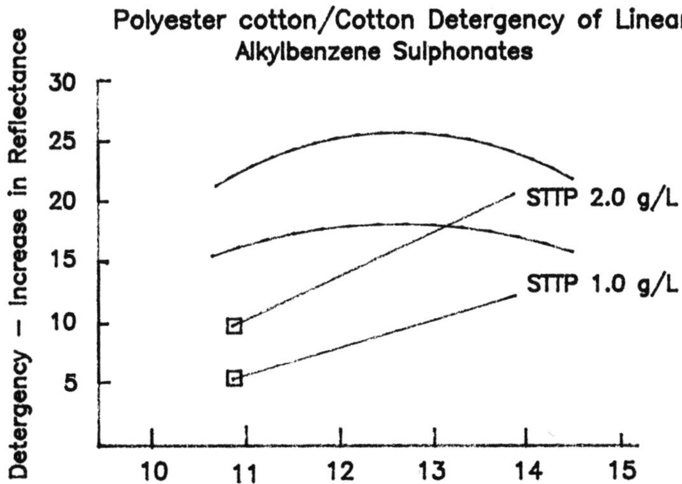

Temperature 60 degree C. Active 0.75 g/L
Water Hardness 300mg/L as calcium carbonate

□ Polyester cotton fabric

The precipitation boundary, consisting of line segments MSL and SPL, divides the graph into regions of clear solutions and cloudy mixtures. Below line segments SML and SPL, only surfactant monomers are present in solution. Above line segments SML and MSL, both micelles and monomers are present. In the region between MSL and SPL, saturated solution exists in equilibrium with the precipitate of the calcium salt of the surfactant.

A detailed explanation and regression equations of the boundary lines are given by Matheson et al[8].

Several methods are available for minimising the effect of this interaction. The most common involves the use of builders to reduce the concentration of water hardness ions. Another is to add more surfactant to make up for the amount lost in the formation of insoluble complexes. A third method involves the use of micelle promotion agents to reduce the extent of interation by effectively lowering the concentrations of both surfactant monomer and water hardness ions.

Figure 9

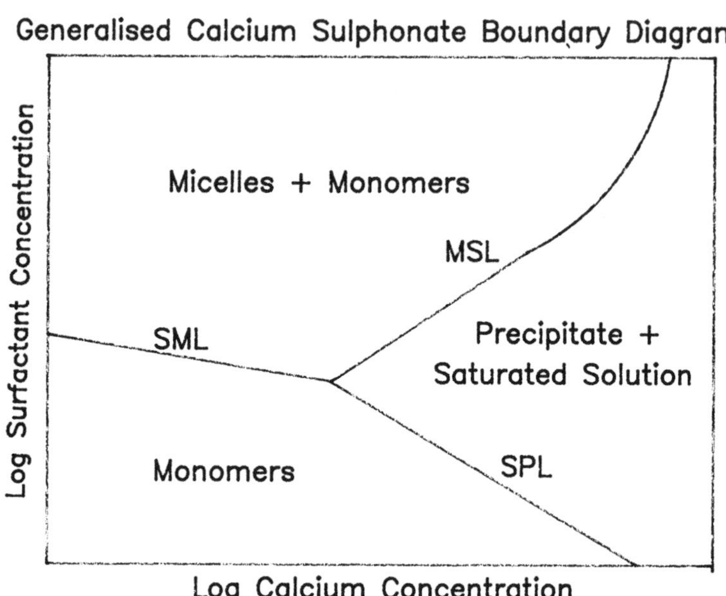

Agents which promote micellisation of linear alkylbenzene sulfonates improve LAS hard water detergency performance by reducing water hardness sensitivity. Inorganic salts and cosurfactants act as micelle promotion agents.

Hard Water Performance

Any agent which promotes micelle formation improves the hard water detergency performance of LAS. Effectively, micelles act as a sink for both surfactant and water hardness ions. Increasing ionic strength is one method of lowering the critical micelle concentration (cmc) of LAS. The addition of sodium sulfate lowers both cmc and surface tension of C_{13}-LAS. These effects result from a decrease in the repulsion of head groups in the micelle and by a reduction in surfactant solubility[9].

The addition of surfactants also can be effective in improving the hard water performance. However, in comparison to increasing ionic strength, the effect of adding a cosurfactant is complicated by interactions which occur between species in solution.

1. LAS and nonionic surfactants interact to produce the nonlinear decrease in cmc observed when NI is substituted for LAS. It is this interaction that produces a substantial reduction in cmc with only a small amount of nonionic surfactant.

2. A greater EO content increases the ability of the surfactant to interact with (and solubilise) Ca $(LAS)_2$.

Environmental and Toxicological Properties of Linear Alkylbenzene Sulphonic Acids and their Salts

Toxicity to fish

Shell laboratory studies on linear alkylbenzene sulphonic acids and their sodium sulphonates have shown that they are, in common with most surface-active agents, toxic to fish at quite low concentrations. Many investigators have established that toxicity decreases as the alkyl chain becomes shorter (Table 5)[10,11].

Table 5 Chronic Toxicity of LAS to Invertebrates and Fish

Species	Chain Length	NEOC mg/l
Invertebrates		
Daphnia magna	C10	9.8
Daphnia magna	C12	4.9
Daphnia magna	C13	0.8
Daphnia magna	C13.3	0.6
Daphnia magna	C14	0.1
Ceriodaphnia sp	C11.7	3.0
Chironomus riparius (water)	C11.7	2.4
Chironomus riparius (sediment)	C11.7	319.0
Gammarus sp	C12	0.2
Mysidopsis bahia	C11.4	0.4
Mysidopsis bahia	C13.1	0.04
Fish		
Fathead Minnow	C10	14.0
Fathead Minnow	C11	7.0
Fathead Minnow	C11.2	5.1
Fathead Minnow	C11.7	1.0
Fathead Minnow	C12	1.0
Fathead Minnow	C13	0.1
Fathead Minnow	C14	0.05
Rainbow Trout	C12	0.12

In practice extensive biodegradation in drains, sewers, waterways and in the course of sewage treatment, always occurs before they come into contact with fish. The toxicity of the biodegraded surface-active agents towards fish is therefore the point of interest in the real environment. Studies have indeed shown that the acute toxicity of linear alkylbenzene sulphonates is reduced by a factor of at least 3-5 in the course of biodegradation[10].

Biodegradability

LAS is probably the most intensively studied molecule in production. Its biodegradational behaviour has been comprehensively studied. Evidence from many workers shows that under aerobic conditions LAS is completely degraded (mineralised) in the environment[1]. The first steps in the mineralisation is accomplished by microorganisms through omega and beta oxidations of the alkyl chain. The resulting sulphocarboxylic acids have a chain length of 4-5 carbon atoms. Splitting of the benzene ring follows and the subsequent desulphonation[13].

In a few laboratory tests incomplete degradation has been obtained but this can be attributed to factors such as concentration and biochemical activity of inoculum, deficiencies and inbalance in media or lack of biomass in synthetic sewage. The removal of LAS in efficiently operated sewage treatment works is high (95-99%). The proportion of LAS associated with sewage sludge is approximately 30%. As LAS is not biodegraded under strict anaerobic conditions it is not removed during anaerobic digestion. However, any LAS present in sludge-amended soils degrades at relatively high rates with half lifes ranging from 3 to 35 days[14].

Based on the concentration in sewage and water usage, the per capita use of LAS in Western Europe is approximately 2.5g/day, which is about 35% lower than the estimate made from total use (556 k.tonne) and population (377 million). The difference can probably partially be explained by the loss of LAS due to biodegradation in the sewage system.

In rivers, concentrations are usually less than 0.05mg/l, although in some countries levels as high as 1 to 2mg LAS/l have been reported where untreated wastewater is discharged. In estuaries and coastal waters the concentrations are very low (c.0.008mg LAS/l) but near outfalls the concentrations can be as high as 0.03mg LAS/l.

LAS is the most widely studied product in commercial production today. By any reasonable standard the degradation of LAS under aerobic conditions is proven.

REFERENCES

1. Reports from The LAS Status Seminar Aachen, 9th November 1988 Tenside 2/89.

2. W.A.A Senden and R. Riemersma, Tenside Surfactants Detergents 1985 22, 319-322.

3. L. Cavalli Presentations to ECOSOL Task Force, Enichem Augusta.

4. Petresa Chemical Company. Sulphonation of L.A.B. with SO_3 Influence of Free Oil (Unsulphonated Matter) and it's Sulphone Content on Physical Properties of Linear Alkylbenzene Sulphonates (L.A.S.), XVIII Journadas CED-AID, Barcelona, March 1987.

5. A. Moreno, J. Bravo, J.L. Berna, JAOCS 1988 65 (6) 1000.

6. A. Moreno, L. Cohn. J.L. Berna, JAOCS 1990 67 (8) 547.

7. K.L. Matheson, M.F. Cox and D.L. Smith, JAOCS 1985 62 (9) 1391.

8. K.L. Matheson, JAOCS 1985 62 (8) 1269.

9. Rosen Milton J., 'Surfactants and Interfacial Phenomena', J. Wiley & Sons, New York 1978.

10. M.A. Lewis , Wat. Res. 1991 25 101-113.

11. R.A. Kimerle , Tenside Surfactants Detergents 1989 26(2) 169-176

12. Brown, Abram and Collins, Tenside Surfactants Detergents 1978 15 57-59.

13. P. Schoberl, Tenside Surfactants Detergents 1989 26(2) 86-94.

14. J. Water M.S. Holt and E. Matthijs, Tenside Surfactants Detergents 1989 26(2) 129-135.

Recent Advances in the Chemistry of Ethoxylated Ethanesulphonate Surfactants

P.K.G. Hodgson and N.J. Stewart

THE BRITISH PETROLEUM COMPANY PLC, RESEARCH CENTRE, SUNBURY ON THAMES, MIDDLESEX TW16 7LN, UK

1 INTRODUCTION

Ethanesulphonates of alkyl/alkylphenylethoxylates have been known since the 1930's. They are more complex to manufacture than conventional surfactants and necessarily command a price premium in the market place. They do possess, however, a unique combination of properties. These are high thermal hydrolytic and chemical stability, high salinity tolerance, a wide range of HLB values through structural modification and high surface activity, one example of which is the ability to develop low interfacial tension in oil-water systems. Over the last decade a number of speciality applications have been reported for ethanesulphonates and these include mineral scale and corrosion inhibition, formulations for emulsion polymerisation and metal/caustic cleaning baths and crude oil recovery.

Our interest in ethanesulphonate surfactants is as components in formulations for Enhanced Oil Recovery (EOR) and injectivity control in deep, offshore oil wells. For such applications the surfactant must be active at low concentration, exhibit low adsorption on to the reservoir, possess hydrolytic stability to high temperature over an extended timescale and be tolerant to a saline environment containing a high concentration of divalent ions at both low and high temperatures.

Initial studies of the physical and chemical properties of a number of series of ethanesulphonates with different hydrophobes and ethoxylate numbers were performed on pure anionic surfactant to enable trends to be more obviously identifiable. It was necessary to establish a rapid method of purification of product mixtures to quantitatively produce large quantities of ethanesulphonate free from nonionic ethoxyalcohol and ethoxychloride precursors. This was achieved with reverse-phase preparative High Performance Liquid Chromatography[1].

The suitability of ethanesulphonates as components of an EOR formulation was assessed by measurement of oil-water interfacial tension and phase volume followed by extent of oil recovery in sand pack and finally core tests. The results of a similar study by the Berol group on commercial, non-purified surfactants have appeared[2]. Thermal hydrolytic stability was determined by ^{13}C NMR spectroscopy in solutions of deoxygenated sea water at pH 6.5. The remarkable stability of the ethanesulphonate class of surfactant was apparent in that less than 5% of degradation was observed after 220 days at 130°C. By contrast, the structurally analogous class of propanesulphonates were considerably less stable (20% degradation, 220 days, 130°C) to carbon-sulphur bond cleavage. These bulk solution stabilities are taken as indicative of stability in reservoir flooding, though other factors such as pressure and catalysts may be relevant there in some circumstances.

This development programme resulted in the identification of a number of highly effective EOR formulations and one of these is detailed in Figure 1[3]. The formulation is a complex mixture of primary and secondary surfactants together with carriers and stabilisers. At 1.5% w/w in sea water it effected removal of 95% of oil in place in a core test at 96°C with oil production in the window 0.7 - 4.5 pore volumes[3].

The economics of surfactant EOR are significantly dependent upon surfactant cost. It was therefore necessary within this programme to evaluate critically the synthetic process for the alkylphenylethoxyethanesulphonate components. This process had changed little to the mid 1980's from the original Rohm and Haas patent[4] of 1938 and is outlined in Figure 2.

A number of areas for improvement in both stages were identified. Drawbacks in the chlorination include long reaction times (which necessitates batch operation), incomplete conversion and ethoxylate chain cleavage which results in less than quantitative yield, toxic 1,4-dioxan removal problems and mismatch of precursor ethoxyalcohol ethoxylate distribution with that of the ethanesulphonate. In the sulphonation stage, long reaction times, non-quantitative conversion, use of excess sulphite, poor reactor volume usage efficiency and viscosity problems, needed to be addressed. The present paper will now review our studies on the optimisation of both synthetic stages.

2 CHLORINATION PROCESS

During conversion of ethoxyalcohols into ethoxysulphonates via ethoxychloride a proportion of the ethoxylate groups is lost. This was established by a comparison of mass spectral distribution of ethoxylate for ethanesulphonate and precursor ethoxyalcohol pairs.

Field application at 1.5% by weight of concentrate in sea water

Polyisobutene Sulphonate (MW 371)	18%
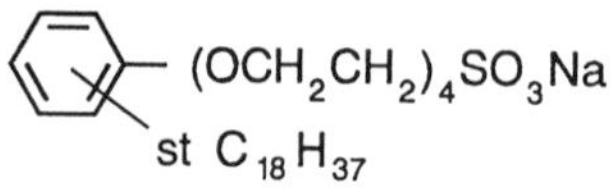 $C_6H_4(\text{st } C_{18}H_{37})-(OCH_2CH_2)_4SO_3Na$	12%
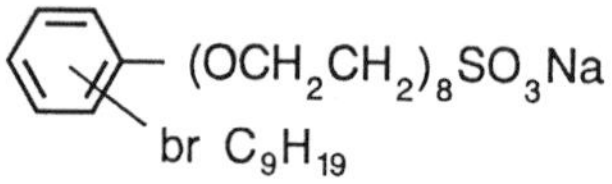 $C_6H_4(\text{br } C_9H_{19})-(OCH_2CH_2)_8SO_3Na$	10%
Kerosene	40%
Water	13%
1 – Hexanol	7%

Figure 1. Enhanced oil recovery surfactant formulation

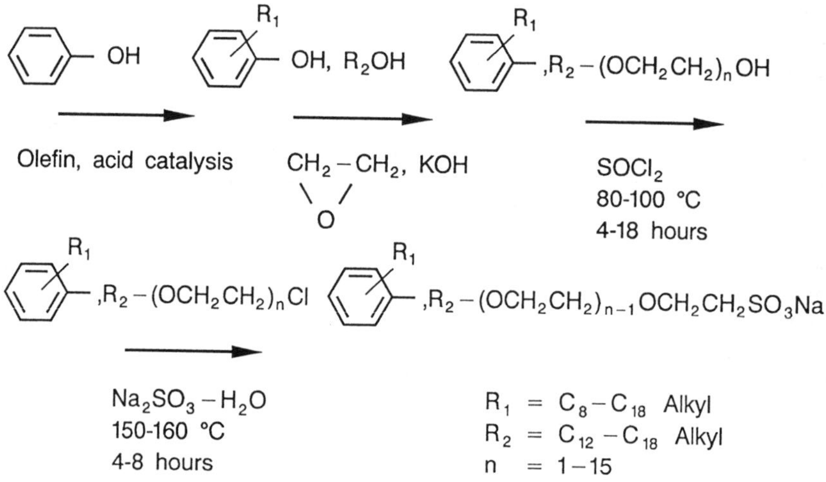

Phenol —(Olefin, acid catalysis)→ R_1-phenol (OH), R_2OH —(CH_2-CH_2 (O), KOH)→ R_1-phenyl, $R_2-(OCH_2CH_2)_nOH$ —($SOCl_2$, 80-100 °C, 4-18 hours)→ R_1-phenyl, $R_2-(OCH_2CH_2)_nCl$ —($Na_2SO_3-H_2O$, 150-160 °C, 4-8 hours)→ R_1-phenyl, $R_2-(OCH_2CH_2)_{n-1}OCH_2CH_2SO_3Na$

$R_1 = C_8 - C_{18}$ Alkyl
$R_2 = C_{12} - C_{18}$ Alkyl
$n = 1 - 15$

Figure 2. Overall manufacturing route

One example is given in Figure 3 for a nominal 9-ethoxyalcohol; an average of two ethoxylate units are lost during the process. This corresponds to a loss of yield of the order of 10% for such high-ethoxylate systems.

Model studies using alkylethoxyalcohols with single fixed ethoxylate chains instead of commercial products with a distribution of ethoxylate chain lengths enabled us to establish that ethoxylate loss occurs exclusively at the chlorination stage and is manifested as 1,4-dioxan production[5]. These findings are summarised in Figure 4. The decomposition mechanism is best explained by oxygen neighbouring group participation in the cleavage of the intermediate chlorosulphite ester. We were unable to detect loss of ethoxylate in the sulphonation process.

The experimental procedures used in this programme have been described[6]. Ethoxylation was performed in a 2 litre stainless-steel autoclave equipped with anchor and turbine agitators by BP Chemicals. Sulphonation was performed in a 1 litre glass Buchi BEP280 autoclave. Product analysis was performed by ^{13}C NMR spectroscopy and depended on a prior analysis of individual components[7].

The model study of Figure 4 indicated that the oxygen-assisted pathway was occurring at a rate similar to that of the unassisted breakdown process. We reasoned therefore, that the assisted 1,4-dioxan pathway could be suppressed and the rate of chlorination increased in a truly catalytic process by the addition of soluble chloride ion. This would shift the mechanism to a direct S_N2 process. Small quantities of Group IA bases/salts were found to be highly effective (Figure 5)[6]. Lithium chloride (1%) reduced chlorination time from 3 hours at 80°C to less than 0.25 hours and suppressed 1,4-dioxan production seventyfold to the limit of detection. Lithium hydroxide was an equally effective catalyst but the hydroxides of sodium and potassium do not catalyse chlorination of this ethoxyalcohol as no rate enhancement and no reduction in 1,4-dioxan production is observed. Interestingly, potassium hydroxide is effective in catalysing the chlorination of more hydrophilic ethoxyalcohols such as those with an increased number of ethoxylate groups. This highlights the importance of a soluble metal halide and, indeed the addition of a crown ether to the potassium hydroxide case of Figure 5 results in a positive catalytic effect. Indeed, some commercial hydrophilic ethoxylates can be rapidly chlorinated through catalysis by residual ethoxylation catalyst.

$C_{18}H_{37}C_6H_4(OCH_2CH_2)_9SO_3Na$
$C_{18}H_{37}C_6H_4(OCH_2CH_2)_9OH$
Percentage
0
20
40
60
80
100
0
2
4
6
8
10
12
14
16
18
20
Number of ethoxylate groups

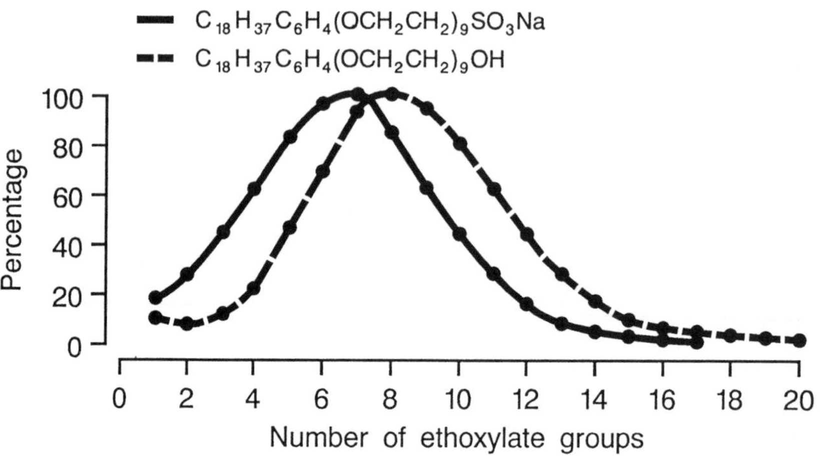

Figure 3. Loss of ethoxylate groups in ethanesulphonate synthesis

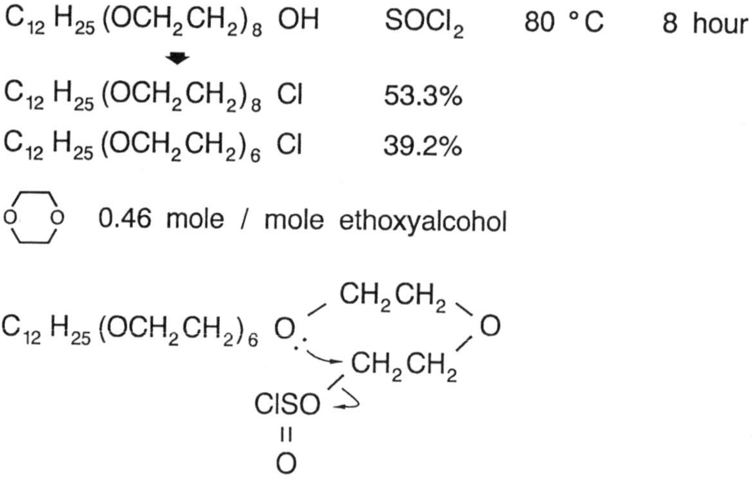

$C_{12}H_{25}(OCH_2CH_2)_8$ OH
$SOCl_2$
80 °C
8 hour
$C_{12}H_{25}(OCH_2CH_2)_8$ Cl
53.3%
$C_{12}H_{25}(OCH_2CH_2)_6$ Cl
39.2%
O
O
0.46 mole / mole ethoxyalcohol
$C_{12}H_{25}(OCH_2CH_2)_6$ O
CH_2CH_2
O
CH_2CH_2
ClSO
O

Figure 4. Model study: ethoxylate groups lost in chlorination stage

The Group IIA bases/salts were all soluble in the ethoxyalcohol of Figure 6 and all effected catalysis. Strontium hydroxide was the most effective catalyst of this series. At 1%, the reaction was complete after 0.5 hours at 80°C and 1,4-dioxan production had been reduced to one-tenth of the uncatalysed level. By contrast, chlorides of iron, aluminium and zinc were ineffective as catalysts and in fact all reduced the chlorination rate and increased 1,4-dioxan production compared to the uncatalysed process. It is important, therefore, to exclude transition and B-metal salts from ethoxyalcohols for chlorination.

This catalytic process for the conversion of alkyl/alkylphenylethoxyalcohols into ethoxychlorides offers a number of economic and processing benefits[8]. Using lithium hydroxide (0.2%), estimated to be the most cost effective catalyst, the resultant twelvefold rate increase would allow the current batch process to be operated at 80°C in a continuous mode with the benefits of increased thoughput and decreased energy requirements. The suppression of 1,4-dioxan production to less than one-tenth of the uncatalysed amount results in increased yields, reduced toxic 1,4-dioxan removal problems and a closer matching of ethoxylate distributions between ethoxyalcohol and ethoxychloride, and hence ethoxysulphonate.

3 ETHOXYLATION/CHLORINATION PROCESS

We realised that certain of the chlorination catalysts were also effective catalysts for the ethoxylation of alcohols and alkylphenols. It is therefore possible to use a single catalyst for both ethoxylation and chlorination stages[9]. This novel concept provides a number of processing advantages. Plant design would be simplified and the cost of employing separate catalysts for each reaction saved. Neutralisation and catalyst removal from the ethoxyalcohol, usually performed to improve storage stability, would be unnecessary. Finally hydrated salts would not require special drying, which would be preferable if they were added at the chlorination stage, and the cheaper metal bases may be employed without waste of chlorinating agent[6].

It is proposed that lithium hydroxide monohydrate (0.2%) be used as the single catalyst for both stages in the conversion of alcohols/alkylphenols into alkyl/alkylphenylethoxychlorides when standard broad-distribution ethoxylates are required. A recent advance in ethoxylation technology is the production of peaked-distribution alkylethoxyalcohols using Group IIA metal bases such as strontium[10]. These newer narrow-distribution ethoxylates are reported to possess improved performance characteristics[11], although it is as yet unknown whether the same will be true of the derived ethanesulphonates. Therefore, it is also proposed that strontium hydroxide (1%) be used as single catalyst for production of narrow-distribution alkylethoxychlorides.

Catalyst	**Chlorination**	**Time at 80 °C**	**Heat-up time**	**1,4-dioxan formation**
(% wt / wt)	(percentage)	(h)	(min)	(moles / mole ethoxyalcohol)
None	88	1	40	0.07 (4 h)
	95	2		
	100	3		
LiCl (1%)	72	0	15	0.001 (2 h)
	100	0.25		
$LiOH.H_2O$ (1%)	100	0	40	0.006 (1.5 h)
$LiOH.H_2O$ (0.2%)	83	0	31	0.006 (1.25 h)
	100	0.25		
KOH (1%)	54	0.25	23	0.06 (3 h)
	78	1		
	92	2		

Figure 5. Chlorination of $C_{18}H_{37}C_6H_4(OCH_2CH_2)_5OH$ with Group IA bases / salts

Catalyst	**Chlorination**	**Time at 80 °C**	**Heat-up time**	**1,4-dioxan formation**
(% wt / wt)	(percentage)	(h)	(min)	(moles / mole ethoxyalcohol)
None	88	1	40	0.07 (4 h)
	95	2		
	100	3		
$CaCl_2$ (1%)	94	0.25	15	0.007 (2.5 h)
	97	0.5		
	100	1		
$Sr(OH)_2$ (1%)	96	0.25	27	0.007 (1.75 h)
	100	0.5		
$FeCl_3$ (1%)	0	0.25	20	1.005 (4 h)
$ZnCl_2$ (1%)	6	0	30	0.37 (4 h)
	23	0.5		
	44	2		

Figure 6. Chlorination of $C_{18}H_{37}C_6H_4(OCH_2CH_2)_5OH$ with Group IIA, transition & B–metal salts / bases

The results in Figure 7 demonstrate how strontium hydroxide effectively catalyses both ethoxylation and chlorination stages with hexadecanol as substrate. Chlorination of as- produced tetraethoxylate with in situ strontium is 90% complete after 2 hours, whereas with the most common ethoxylation catalyst, potassium hydroxide, chlorination, which is uncatalysed, is only 26% complete. 1,4-Dioxan production is reduced tenfold in the strontium catalysed chlorination.

4 SULPHONATION PROCESS

A typical reaction profile for sulphonation of an ethanechloride by reaction with an excess of sodium sulphite in water is depicted in Figure 8. The initial rate of sulphonation is low until sufficient surfactant sulphonate is produced to emulsify the two reaction phases. Sulphonation is then reasonably rapid. The induction period can be essentially removed by addition of 2-10% of final product sulphonate[12] so that the starting point is now the vertical broken line on Figure 8.

Even so, scope remains to reduce reaction time. It has proved impossible to achieve conversions to ethanesulphonate above 80-90% due to competing hydrolysis of ethanechloride to ethoxyalcohol. This side-reaction appears to be pronounced in the later stages of the reaction; the sulphonation level reaches a maximum before all of the ethanechloride is consumed. It has also proved to be difficult with this system to reduce water content in the reactor below 40% of charge and to reduce sodium sulphite level to below 1.3 equivalents. In addition, gel formation is noted during sulphonation of the more hydrophobic ethanechlorides which markedly slows reaction rate.

It was perceived that the drawback with using sodium sulphite is that its solubility in water is low and decreases at higher temperature and is considerably less than sodium chloride into which it is converted during the reaction. This dictates that substantial quantities of water must be present to maintain solubility and that plateau formation of ethanesulphonate might be related to salting out of sulphite. By contrast, potassium sulphite is much more soluble than sodium sulphite, has increasing solubility with temperature and has a higher solubility than potassium chloride.

Potassium sulphite is a more effective reagent than sodium sulphite in Strecker sulphonation and this is demonstrated by the reaction profiles in Figure 9[13]. This relatively hydrophobic ethanechloride is difficult to sulphonate efficiently using sodium sulphite and only 73% conversion was achieved after 6 hours. This value rose to 80%, however, when equimolar potassium sulphite replaced the sodium sulphite. Gelation was observed after 3.5 hours in each case. An even more significant improvement may yet be achieved with potassium sulphite through reduction of water content.

Catalyst		**Ethoxylation** (4 equivalent ethylene oxide)	**Chlorination** (1.08 equivalent thionyl chloride 80 °C)
$Sr(OH)_2$	(1%)	125 °C 4 bar 5 h 1.5% phenol promoter	90% conversion 2 h 1.4-dioxan 0.02 M/M
KOH	(0.3%)	125 °C 4 bar 5 h	26% conversion 2 h 1.4-dioxan 0.2 M/M

Figure 7. Catalysis of ethoxylation & chlorination of hexadecanol

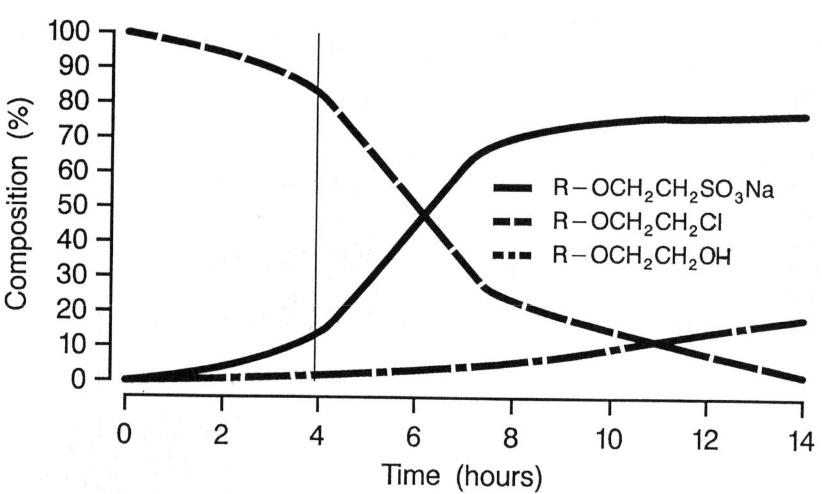

Figure 8. Typical sodium sulphite-based Strecker reaction profile

With sodium sulphite, conversion and reaction rates are substantially reduced at lower water levels. With potassium sulphite by contrast, on the basis of solubility arguments, it is now possible to reduce water content(and so improve reactor usage efficiency) to very low levels without reduction in reaction rate or conversion to ethanesulphonate. In fact because sulphite ion concentration has now been increased, sulphonation rate is actually increased and extent of hydrolysis reduced (Figure 10). With the lowest water content (down to 13% of original), conversion to sulphonate as high as 89% after 6 hours was achieved and, for the first time with this system, gelling was avoided. Reactor volume use efficiency was increased from 35% to 60% ethanechloride content.

The sulphonation of an ethoxyethanechloride formed from a narrow-range ethoxylate distribution ethoxyalcohol had not been reported and it was therefore of interest to apply our procedures to such a substrate. The results of a direct comparison between broad- and narrow-ethoxylate hexadecanol 4-ethoxyalcohols are shown in Figure 11[4].

The narrow-range ethoxyalcohol was chlorinated rapidly through catalysis by in situ soluble strontium chloride. Chlorination to completion of the broad-range ethoxyalcohol required much longer as the insoluble potassium chloride is not catalytic for this substrate because of the absence of a crown ether type solubilising effect.

Sulphonation was performed with 3 equivalents of sulphite in aqueous ethanol at 155-160°C for 6 hours[14]. With the broad-range ethoxychloride and sodium sulphite, 65% conversion to sulphonate was achieved with 35% hydrolysis. This is a typical result for such hydrophobic ethoxychlorides. With the narrow-range ethoxychloride, sulphonation was increased to 80%. We believe this improvement to be at least partly a result of a lower proportion of the more hydrophobic shorter ethoxylate chain components in the narrow-range product which are more difficult to sulphonate. Application of potassium sulphite to the narrow-range ethoxychloride dramatically increased conversion to quantitative. It remains to be established, however, for a particular application what performance improvement has been achieved with the high-conversion, narrow-range ethanesulphonate products.

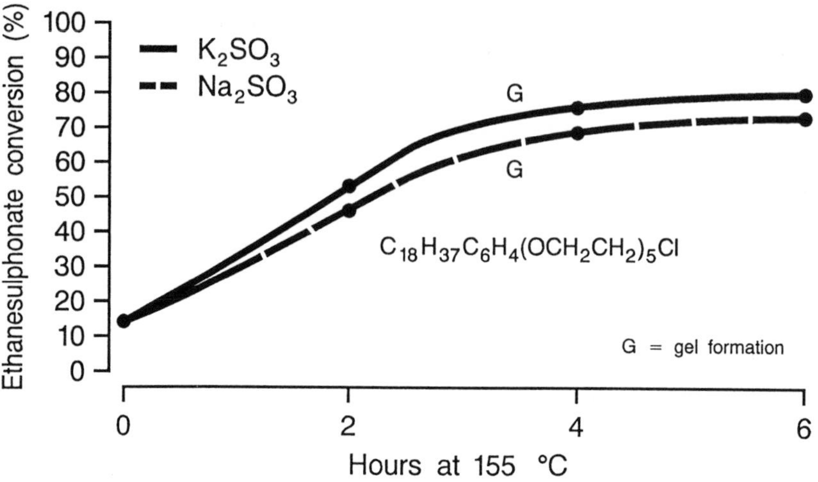

Figure 9. Sulphonation at equivalent Na_2SO_3 & K_2SO_3 concentrations

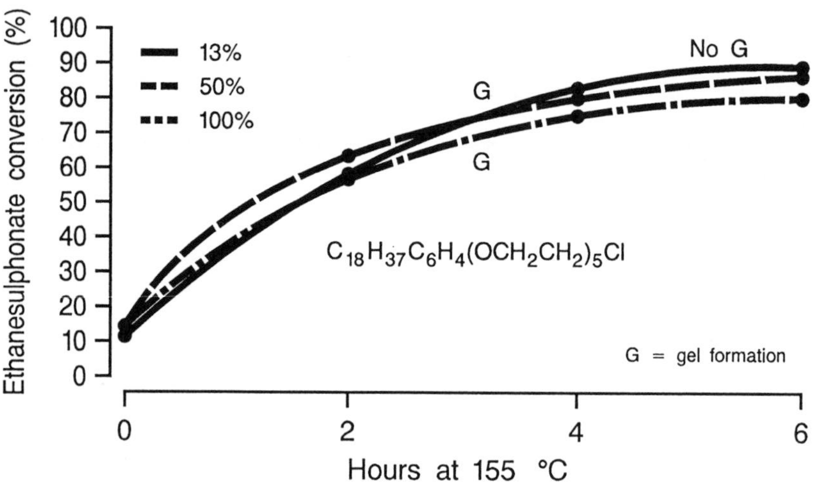

Figure 10. Sulphonation with K_2SO_3 at 100%, 50% & 13% of original water content

Ethoxylate distribution	**Chlorination time to completion**[a] (hours)	**Sulphonation ratio $CH_2SO_3M:CH_2OH$**[a]
Broad	8	65:35[b]
Narrow	2	80:20[b]
Narrow	2	100:00[c]

a Determined from ^{13}C NMR spectrum

b Sulphonation with Na_2SO_3

c Sulphonation with K_2SO_3

Figure 11. Chlorination & sulphonation of broad & narrow $C_{16}H_{33}(OCH_2CH_2)_4OH$

5 CONCLUSIONS

* Ethoxylated ethanesulphonate surfactants are a versatile class of speciality surfactant of high thermal hydrolytic stability.

* Chlorination of ethoxyalcohols is catalysed by metal salts/bases; ethoxylate cleavage is suppressed.

* A single catalyst can be used for both ethoxylation and chlorination - lithium for broad-distribution and strontium for narrow-distribution ethoxylates.

* Potassium sulphite is a more effective reagent than sodium sulphite in the sulphonation process - reduced reaction times and higher reactor volume efficiency are achieved.

* Narrow-range ethoxychlorides are sulphonated to higher conversion than broad-range ethoxychlorides. Replacement of sodium sulphite by potassium sulphite increases conversion still further.

6 REFERENCES

1. P K G Hodgson and N J Stewart, J Chromatography, 1987, 387, 546.
2. K Holmberg and E Oesterberg, Proceedings of the Third European Symposium on Enhanced Oil Recovery, Rome, 1985, 2, 211.
3. M A Lee, J C Morgan and A Pearce (British Petroleum), UK Patent 2 158 126 (1985).
4. H A Bruson (Rohm and Haas), US Patent 2 115 192 (1938).
5. P K G Hodgson, N J Stewart and E J Tinley, Tenside Detergents, 1986, 23, 175.
6. P K G Hodgson, and N J Stewart, Applied Catalysis, 1989, 46, 313.
7. J D Boyle (British Petroleum, Sunbury Research Centre, Analytical Division), unpublished results;
   P E Figdore, Org Mag Res, 1983, 21, 678.
8. P K G Hodgson and N J Stewart (British Petroleum), EP 186 451 (1986).
9. P K G Hodgson and N J Stewart (British Petroleum), EP 185 554 (1986).
10. K Yang, G L Nield and P H Washecheck (Conoco), US Patent 4 223 164 (1980).
11. K L Matheson, T P Matson and K Yang, J Amer Oil Chem Soc, 1986, 63, 365; K W Dillan, ibid, 1985, 62, 1144.
12. W A Lidy (BP Chemicals), EP 156 601 (1985).
13. N J Stewart, Proceedings of Second World Surfactants Congress, Paris, 1988, 2, 4;
    N J Stewart (British Petroleum), EP 206 678 (1986).
14. P K G Hodgson, N J Stewart, C E Grant and A M Nicolls, J Amer Oil Chem Soc, 1990, 67, 730.

Applications of *N*-acyl Sarcosine Surfactants

J.J. Crudden, B.J. Lambert and R.W. Kohl

TECHNICAL SERVICE LABORATORY, W.R. GRACE & CO.-CONN., POISSON AVENUE, NASHUA, NH 03060, USA

Introduction:

N-acyl sarcosines and their salts, the sarcosinates, are mild biodegradable anionic surfactants produced from fatty acids and the amino acid sarcosine. They exhibit maximum surfactancy at mildly acidic pH's reaching surface tensions as low as 21 dynes cm^{-1} and are substantive by adsorption of the head group to many materials such as metals, glass and proteinaceous surfaces. The unique properties of these surfactants have led to their use in areas as diverse as toothpaste and shampoo formulation, ore flotation, corrosion inhibition and injection molding.

Synthesis:

Sarcosine, N-methyl glycine, is a naturally occurring amino acid. It can be produced by thermal decomposition of caffeine in the presence of barium carbonate. Sodium sarcosinate is manufactured commercially from formaldehyde, sodium cyanide and methyl amine.

N-acyl sarcosines are condensation products of sodium sarcosinate and a fatty acid chloride (Figure 1). The fatty acid chloride is reacted with sodium sarcosinate under alkaline conditions to produce the fatty sarcosinate sodium salt which is water soluble. Upon acidification, the fatty sarcosine acid, which is water insoluble, is formed and may be isolated from the reaction medium. The acyl sarcosines may be neutralized with bases such as Na, K, NH_3, or TEA to produce aqueous solutions.

The specifications for some commercially available forms of N-acyl sarcosine are presented in Table 1. The products are typically 94% active and may contain up to 6% free fatty acid. Lauroyl and myristoyl sarcosine are produced from refined C_{12} and C_{14} fatty acids. Cocoyl sarcosine is produced from unrefined coconut fatty acid. Cocoyl and oleoyl sarcosine are liquids at room temperature, the others are white waxy solids which melt between 35 and 50°C.

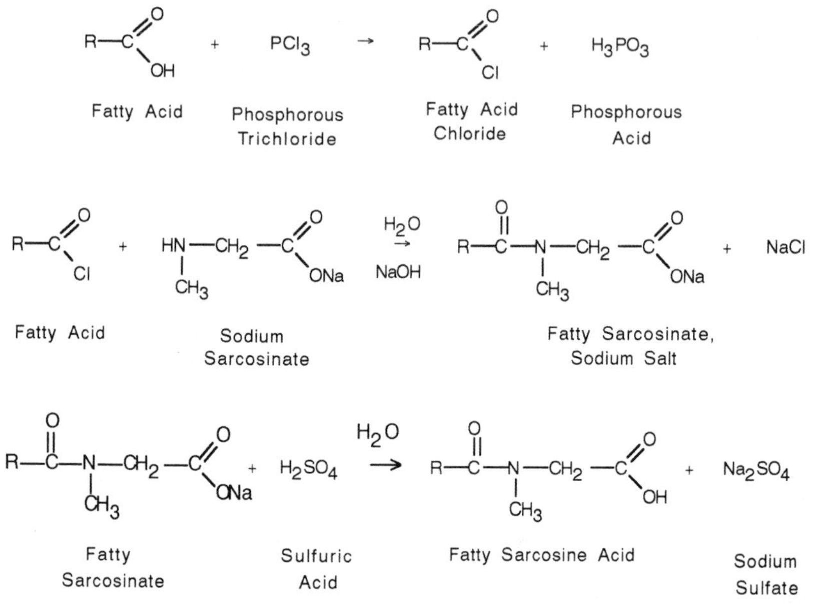

$$R{-}C(=O){-}OH \; + \; PCl_3 \; \rightarrow \; R{-}C(=O){-}Cl \; + \; H_3PO_3$$

Fatty Acid | Phosphorous Trichloride | Fatty Acid Chloride | Phosphorous Acid

$$R{-}C(=O){-}Cl \; + \; HN(CH_3){-}CH_2{-}C(=O){-}ONa \; \xrightarrow[NaOH]{H_2O} \; R{-}C(=O){-}N(CH_3){-}CH_2{-}C(=O){-}ONa \; + \; NaCl$$

Fatty Acid | Sodium Sarcosinate | Fatty Sarcosinate, Sodium Salt

$$R{-}C(=O){-}N(CH_3){-}CH_2{-}C(=O){-}ONa \; + \; H_2SO_4 \; \xrightarrow{H_2O} \; R{-}C(=O){-}N(CH_3){-}CH_2{-}C(=O){-}OH \; + \; Na_2SO_4$$

Fatty Sarcosinate | Sulfuric Acid | Fatty Sarcosine Acid | Sodium Sulfate

Figure 1 Synthesis of Acyl Sarcosines

Acyl Sarcosines may be neutralized with Na, K, NH_3 or TEA to produce aqueous solutions.

Table 1

Specifications for Commercially Available Acyl Sarcosines

	Lauroyl Sarcosine	Cocoyl Sarcosine	Myristoyl Sarcosine	Oleoyl Sarcosine	Stearoyl Sarcosine
Active Ingredient	94% min.	94% min.	94% min.	94% min.	94% min.
Free Fatty Acid	6% max.	6% max.	6% max.	6% max.	6% max.
Color, Gardner	2 max.	3 max.	2 max.	4 max.	4 max.
Appearance @ 25°C	white waxy solid	pale yellow liquid	white waxy solid	yellow liquid	white waxy solid

The specifications for some commercially available N-acyl sarcosinate sodium salts are presented in Table 2. The sodium salts which are water soluble are typically supplied as 30% solutions in water. Sodium lauroyl sarcosinate powder, produced by spray drying the 30% solution, is also available.

Table 2

Specifications for Commercially Available Sarcosinate Salts

	Sodium Lauroyl Sarcosinate Solution	Sodium Lauroyl Sarcosinate Powder	Sodium Cocoyl Sarcosinate Solution	Sodium Myristoyl Sarcosinate Solution
Active Ingredient	30%±1%	95% min.	30%±1%	30%±1%
pH, 10% solution	7.5 - 8.5	7.5 - 8.5	7.5 - 8.5	7.5 - 8.5
Sodium Soap	2% max.	4% max.	2% max.	2% max.
Color, APHA, as is	60 max.	80 max. (20% solution)	100 max.	80 max.
Appearance	colorless liquid	dry white powder	colorless to very pale yellow liquid	colorless liquid

The fatty acid chain length distributions of these commercially available forms are presented in Table 3.

Table 3

Typical % Fatty Acid Chain Distribution in Sarcosinates

	Saturated							**Unsaturated**			
	C_6	C_8	C_{10}	C_{12}	C_{14}	C_{16}	C_{18}	$1{=}C_{16}$	$1{=}C_{18}$	$2{=}C_{18}$	$3{=}C_{18}$
Cocoyl	1.0	7.0	6.0	49.0	18.0	8.0	6.0		5.0		
Lauroyl			0.5	99.0	0.5						
Myristoyl				1.0	98.0	1.0					
Oleoyl	[1.0 ←			10.0		→	1.0]	6.0	71.0	10.0	2.0

Surface Properties:

The CMC and minimum surface tension of several N-acyl sarcosinates at pH 6.8 are presented in Table 4. The data for sodium lauryl sulfate is presented for comparison.

Table 4

Surface Properties

	CMC, w/w%	Minimum Surface Tension dynes/cm
Sodium Lauroyl Sarcosinate	8.0×10^{-2}	24.3
Sodium Myristoyl Sarcosinate	7.9×10^{-3}	27.2
Sodium Cocoyl Sarcosinate	8.7×10^{-3}	22.7
Sodium Oleoyl Sarcosinate	2.6×10^{-3}	28.0
Sodium Lauryl Sulfate	2.4×10^{-1}	33.5

pH Dependence of Surface Tension:

The pH dependence of surface tension for some sarcosinates, above their CMC's is presented in Figure 2. The surface tension is highest above pH 9 and decreases to a minimum for all these surfactants between pH 6 and 7. The cocoyl sarcosinate shows a deep minimum at pH 7. Below pH 5 the surfactants are converted predominantly to the acid form and begin to separate from solution.

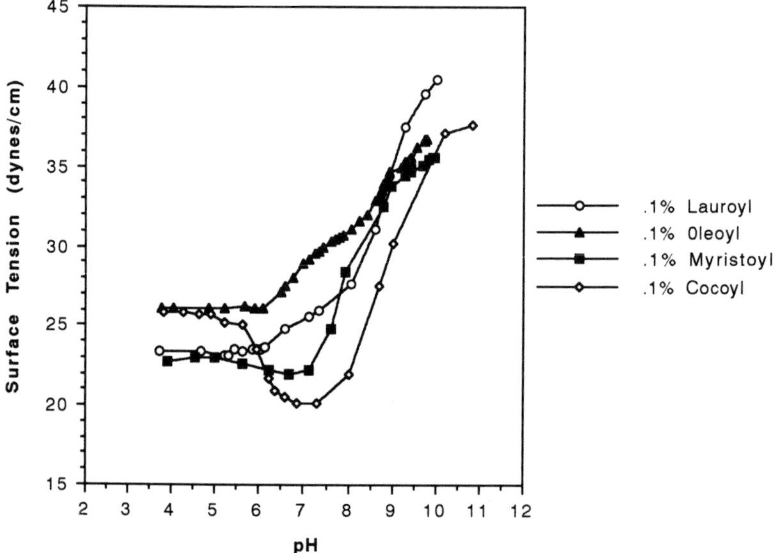

Figure 2 N-Acyl Sarcosinates pH Dependence of Surface Tension

Hydrophile Lipophile Balance (HLB) Numbers:

The HLB number of an emulsifier is an expression of the balance of size and strength of hydrophilic and lipophilic groups of an emulsifier. Emulsifiers of predominantly lipophilic character are assigned numbers below 9 while those of predominantly hydrophilic character are assigned numbers above 11. The HLB system was developed by Griffin (1).

Davies (2) developed a method of calculating HLB values for ionic surfactants by assigning HLB numbers to individual functional groups in the molecule. HLB numbers for some N-acyl sarcosine acids and their sodium salts, calculated using the method of Davies, are presented in Table 5. The numbers for the acids range from 9.5 to 13, indicating that they should be useful as emulsifiers and wetting agents and as additives in hydrocarbon systems such as fuels and lubricants. The HLB numbers for the sodium salts range from about 26.5 to 30, indicating that they should be effective detergents and solubilizers in aqueous systems.

Table 5

HLB Values

ACIDS	HLB
Lauroyl Sarcosine	13.1
Myristoyl Sarcosine	12.1
Cocoyl Sarcosine	10±0.5
Oleoyl Sarcosine	9.6
SODIUM SALTS	
Sodium Lauroyl Sarcosinate	29.8
Sodium Myristoyl Sarcosinate	28.9
Sodium Cocoyl Sarcosinate	27±0.5
Sodium Oleoyl Sarcosinate	26.6
Sodium Lauryl Sulfate	40.0

Substantivity of Sarcosinates:

The sarcosinates adsorb strongly on the surfaces of many substrates such as glass, ceramics, metals and proteinaceous surfaces.

Nelson and Stewart (3) showed, using C_{14} isotopic labeling, that sarcosinates adsorb strongly onto casein, gelatin and human hair. The adsorption is enhanced at lower pH and higher concentration. They also found that the sarcosinate adsorbed to a greater extent on damaged than on undamaged hair.

Studies by Salensky et al (4) suggest that when sarcosinates adsorb on metals, surface chelation occurs through the amino acid portion of the molecule. They found that the carboxylate moiety orients itself almost vertically, and concluded that the surface metal is chelated in the form of five membered ring as depicted in Figure 3.

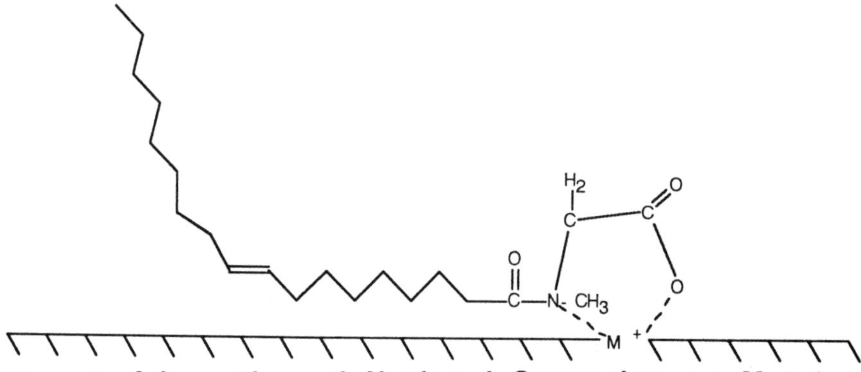

Figure 3 Adsorption of N-oleoyl Sarcosine on Metal

Bettley [5], Figure 4, showed by a diffusion cell experiment that sodium lauroyl sarcosinate actually reduces the permeability of human epidermis to water and other solutes, presumably by depositing a hydrophobic protective layer on the surface.

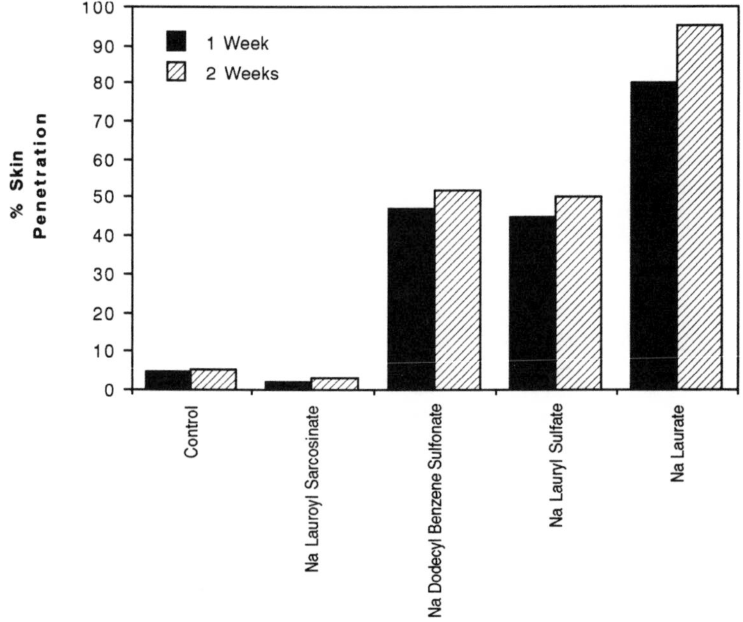

Figure 4 Epidermal Permeability At 0.04M

In a patent assigned to Diversey Corporation, Dvorkovitz et al [6], claim that sodium lauroyl sarcosinate reduces the irritation caused by components such as soda ash and tetrapotassium pyrophosphate in detergent formulations.

Contact Angle:

The contact tangent angle formed between a "sessile" drop and it's supporting surface is directly related to the forces at the liquid solid interface.

The variation of advancing contact angle between sodium lauroyl sarcosinate solution and steel alloy J55 with pH and concentration is presented in Figure 5. The contact angle decreases with decreasing pH and increasing concentration.

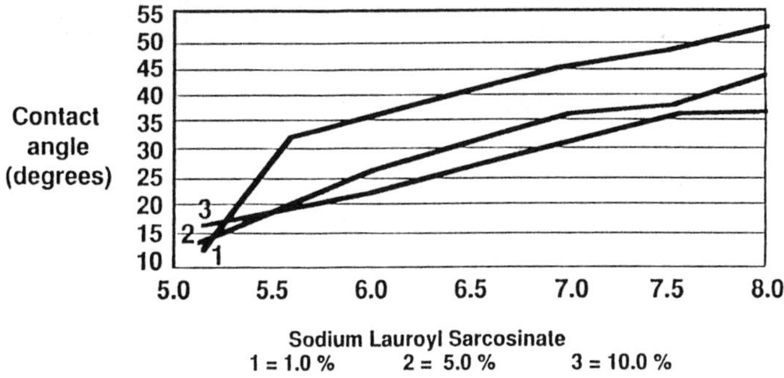

Figure 5 **Contact Angle on Steel Alloy J55**
Dependence on pH

Foaming and Lathering:

The sarcosinates are effective foaming and lathering agents. The lather drainage times for a range of N-acyl sarcosinates determined by the method of Hart and DeGeorge [7] are presented in Figure 6. The lauroyl sarcosinate can be seen to be the most effective lathering agent.

Sarcosinates foam well in electrolyte solutions. The foam volume of sodium lauroyl sarcosinate as a function of sodium chloride concentration is presented in Figure 7. Foam volume reaches a maximum at about 8% sodium chloride. The foam volume of sodium dodecyl sulfate, by contrast, is depressed by any addition of sodium chloride.

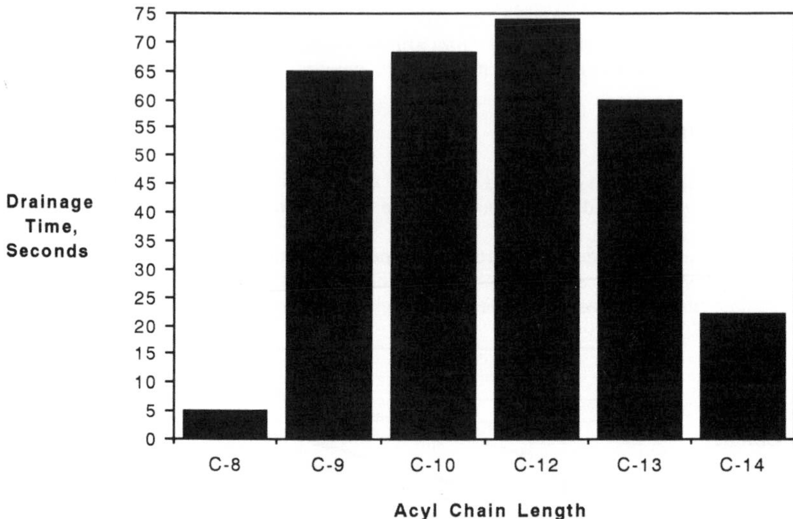

Figure 6 Lather Drainage Tests of Sarcosinates At 1% active ingredient, pH 6.5

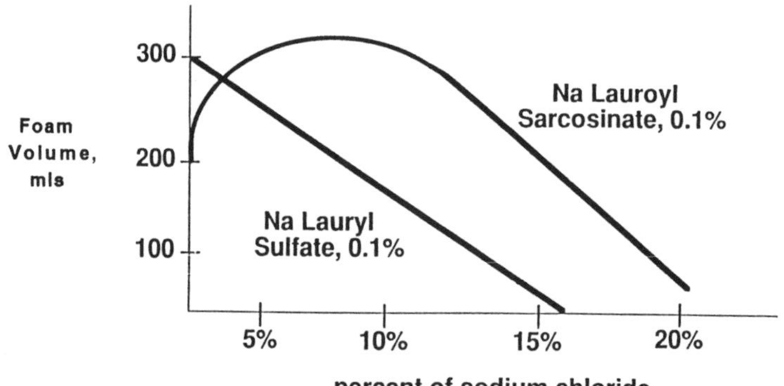

Figure 7 Effect of Sodium Chloride on Foaming

The lather drainage time for sodium lauroyl sarcosinate is increased by the presence of hardness up to about 100 ppm calcium carbonate, as shown in Figure 8, and by the presence of synthetic sebum.

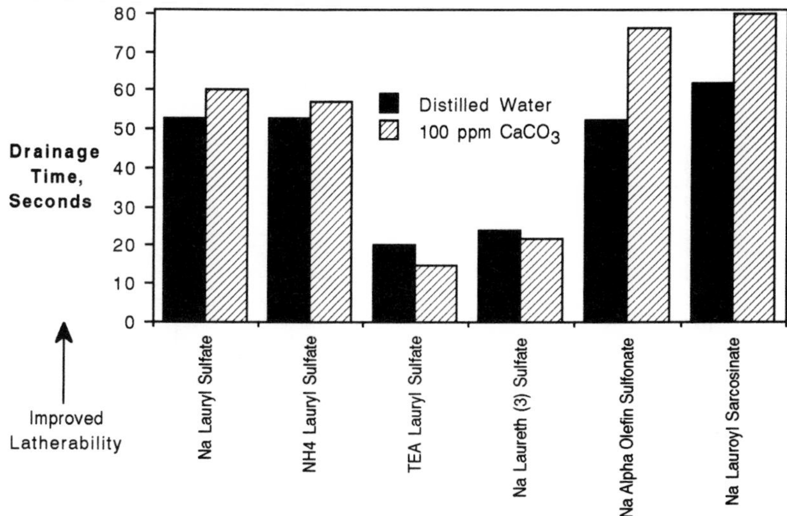

Figure 8 Effect of Water Hardness on Lather Drainage Times of Various Surfactants Concentration 0.80 (w/w) %

Viscosity Control:

A 15% solution of sodium myristoyl sarcosinate, (Figure 9), exhibits increased viscosity below pH 6 in the presence of 1% sodium chloride. This effect can be used to build viscosity in mixed surfactant systems. As the solids content is increased the effect moves to higher pH.

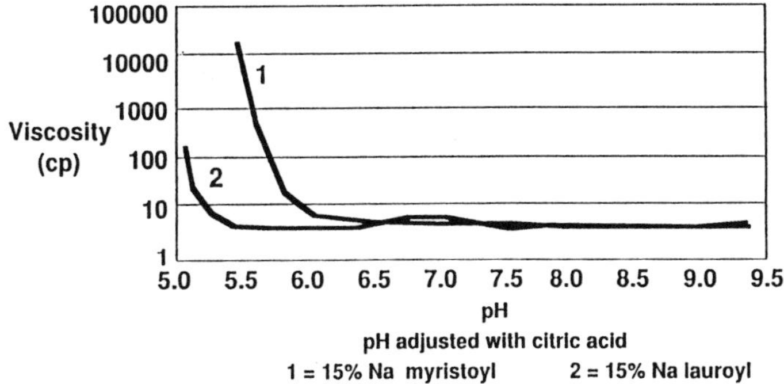

Figure 9 Dependence of Viscosity on pH
Sarcosinate Solutions Containing 1% NaCl

A polyol alkoxy ester - Crothix™* (Figure 10) has been found to be an effective viscosity builder in sodium lauroyl sarcosinate at pH 7.

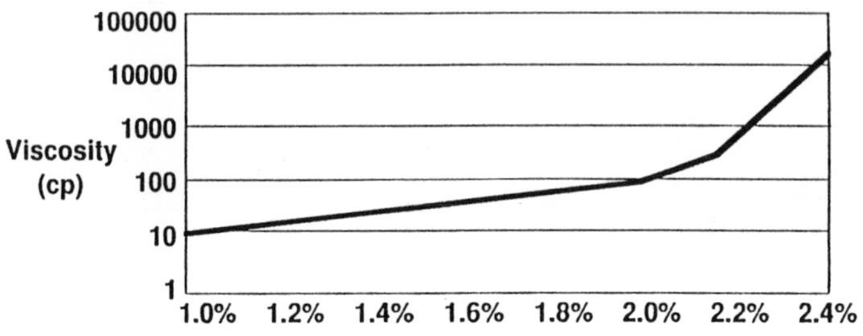

Figure 10 Effect of Polyol Alkoxy Ester on the Viscosity of 10% Sodium Lauroyl Sarcosinate

Sodium lauroyl sarcosinate has been found useful for reducing viscosity in some high solids systems.

Environmental Properties:

Biodegradability:

Biodegradation rates of various N-acyl sarcosines, determined over a 41 day period, using the biochemical oxygen demand protocol, published by the American Public Health Association, along with the figures for sodium lauryl sulfate for comparison are presented in Table 6. All the materials tested are readily biodegradable.

Table 6

Biodegradability of Acyl Sarcosinates

Biological oxygen demand of the surface active agents after a 41 day incubation period.**

Surface-Active Agent	% COD	% of Theoretical Oxygen Demand
Lauroyl Sarcosine	83	78
Cocoyl Sarcosine	76	74
Oleoyl Sarcosine	76	70
Sodium Lauryl Sulfate	65	67

*Croda, Inc.
**A.P.H.A. Protocol

Acute Oral Toxicity:

The acute oral toxicity of some sarcosines and sarcosinates in rats and mice is presented in Table 7. The figures range from 2,000 to 6,000 mg/kg. The LD_{50} of sodium dodecyl sulfate in rats, by comparison, is 1288 mg/kg.

Table 7

Acute Oral Toxicity

PRODUCT	**LD_{50}, mg/kg**	**REMARKS**
Sodium Lauroyl Sarcosinate	2175, 5000	mice, rats
Sodium Cocoyl Sarcosinate	4200	rats
Cocoyl Sarcosine	5400	rats
Sodium Oleoyl Sarcosinate	6000	rats
Lauroyl Iminodiacetic Acid	2600, 4200	rat, females rat, males

Aquatic Toxicity:

The aquatic toxicity of lauroyl sarcosinate to fish is significantly less than other anionic surfactants. Using the protocols described in ASTM Designation E729-88a it was determined that the 96 hour LC_{50} of lauroyl sarcosine for rainbow trout is 56 mg active ingredient per liter (Table 8). The aquatic toxicity of sodium dodecyl sulfate generally ranges from 1-10 mg per liter.

Table 8

Fish Toxicity
Rainbow Trout

Lauroyl Sarcosine

24 hour LC_{50}	93 mg/liter
48 hour LC_{50}	61 mg/liter
72 hour LC_{50}	56 mg/liter
96 hour LC_{50}	56 mg/liter

100% dead at 72 mg/liter
100% survived at 46 mg/liter

LC_{50} for Sodium Lauryl Sulfate is typically 1-10 mg/liter.

Bactericidal Effects:

The bactericidal effects of the sarcosinates are most pronounced under acidic conditions. They have good activity against straphylococcus aureus below pH 5.8 and pseudomonas aeruginosa below pH 4. The bactericidal flora of human saliva are inhibited by sarcosinates at concentrations as low as 0.25%.

Unlike most other anionic surfactants, sarcosinates are compatible with cationic germicides, usually without loss of activity. A list of germicides which are compatible with sarcosinates are presented in Table 9.

Table 9

Antibacterial Compounds that are Compatible with Acyl Sarcosine Surfactants

	Quaternary Ammonium Compounds	
Benzalkonium Chloride	Lauralkonium Chloride	Quaternium 5
Benzelhonium Chloride	Laurtrimonium Chloride	Quaternium 12
Cetalkonium Chloride	Methylbenzethonium Chloride	Quaternium 14
Cetrimonium Chloride and Bromide	Myristalkonium Chloride	Quaternium 17
	Myrtrimonium Bromide	Stearalkonium Chloride
Cetylpyridinium Chloride	Quaternium 2	Steartrimonium Chloride
	Phenolics	
o-Benzyl-p-Chlorophenol	2,4-Dichlorophenol	Resorcinol
Coal tar	Hexachlorphene	Salicylic Acid
o-Chlorophenol	Hexylresorcinol	Thymol
p-Chlorophenol	Pentachlorophenol	2,4,5-Trichlorophenol
Chlorothylmol	Phenol	Tribromosalan
Chloroxylenol (PCMX)	o-Phenylphenol	Triclosan
Dichlorophene		
	Others	
Benzoyl Peroxide	Chlorhexidine and Salts	Sulfur
Benzyl Alcohol	Retinoic Acid	Zinc Pyrithione
2,4-Dichlorobenzyl Alcohol		

The Use of Sarcosinates to Enhance the Performance of Other Surfactants:

Sarcosinates are compatible with nonionic, anionic and cationic surfactants and can often be used to enhance the performance of other surfactants or surfactant systems.

Krafft Point Depression:

The Krafft point of an anionic surfactant solution is the temperature below which the surfactant falls out of solution at concentrations above the CMC.

Below the Krafft point the surfactant is incapable of solubilization and formulations become unstable.

N-acyl sarcosinates have been found to be very effective at depressing the Krafft point of anionic surfactants. The effect of sodium lauroyl sarcosinate on the Krafft point of sodium dodecyl sulfate and sodium cocoyl isethionate is presented in Figure 11. The addition of 0.5% by weight of sarcosinate to the SDS depresses the Krafft point to 0°C. Addition of 1% by weight of sarcosinate to the SCI depresses the Krafft point from 17°C to 0°C.

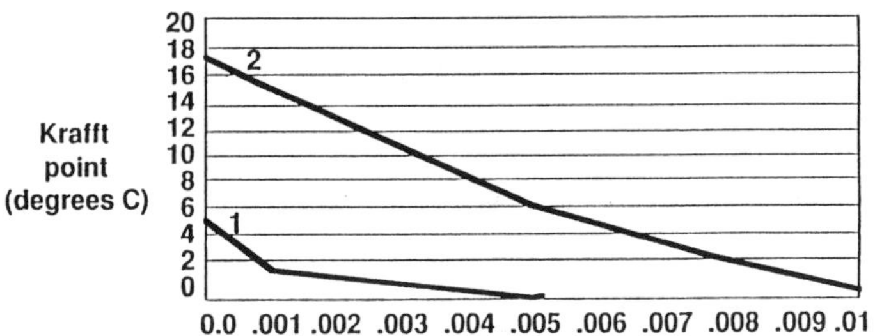

Figure 11 Krafft Point Depression of 1.0% Surfactant Solutions

Elevation of Cloud Point:

As the temperature of an aqueous solution of some nonionic surfactants is raised, a temperature is reached at which the solution begins to become cloudy as it begins to separate into two phases. This temperature is referred to as the cloud point.

It is sometimes advantageous to adjust the cloud point of a surfactant in relation to the expected use temperature of the formulation. At the cloud point, foam formation is reduced while detergency begins to be enhanced, reaching a maximum at 15°C or more above the cloud point [(8)].

Sarcosinates have been found useful for elevating the cloud point of nonionic surfactants. The effect of sodium lauroyl sarcosinate on the cloud point of some commercially available nonionic surfactants is presented in Figure 12. The addition of sarcosinate raises the cloud point in each instance, allowing it to be set to a specific required value.

Depression of Surface Tension:

Sarcosinates have been found, in some instances, to depress the minimum surface tension achievable with other surfactants when incorporated as co-surfactants.

The surface tension of a 1% solution of surfactant varying from pure sucrose cocoate to pure lauroyl sarcosinate at neutral and acidic pH is presented in Figure 13. At 20% sarcosinate the surface tension is 5 dynes cm^{-1} lower than that achieved with pure sucrose cocoate.

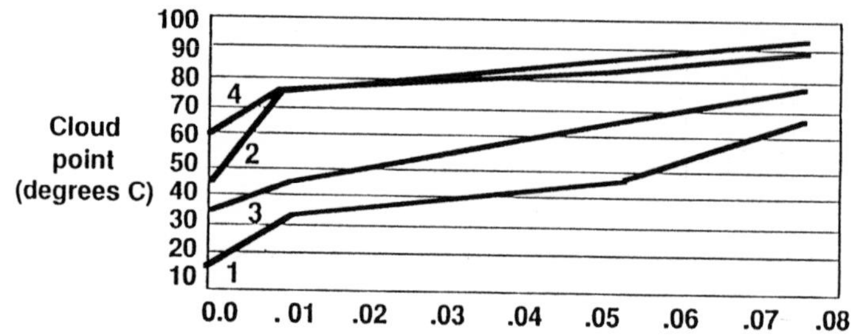

% Active Sodium Lauroyl Sarcosinate

1=Alcodet MC-2000 2=Siponic NP-9
3=Tergitol 15-S-7 4=Tergitol 15-S-9

Tergitol™, Union Carbide Corp.
Alcodet™, Alcolac
Siponic™, Alcolac

Figure 12 **Cloud Point Elevation**
(All Solutions are 1.0% Active Surfactant)

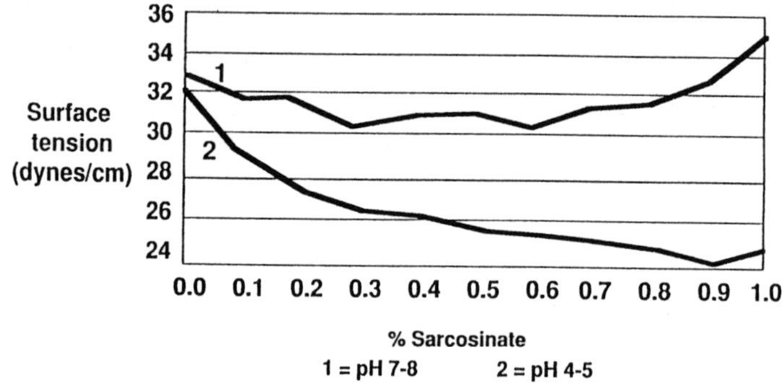

% Sarcosinate
1 = pH 7-8 2 = pH 4-5

Figure 13 Na Lauroyl Sarcosinate : Sucrose Cocoate
Surface Tension vs. Composition (1% Solution)

A similar study carried out using lauryl dimethyl amine oxide with sodium lauroyl sarcosinate is presented in Figure 14. A 30% addition of sarcosinate reduces the surface tension of the system by 5 dynes under both neutral and acidic conditions, and an addition of 10% lauryl dimethyl amine oxide to the sarcosinate reduces the surface tension of the system by about 5 dynes under neutral conditions.

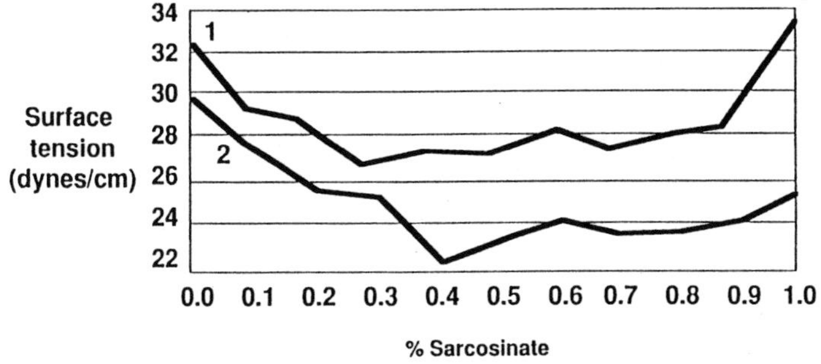

Figure 14 Na Lauroyl Sarcosinate: Lauryl Dimethyl Amine Oxide Surface Tension vs. Composition (1% Solution)

The concentration dependence of surface tension of mixtures of sodium dodecyl sulfate (SDS) and sodium lauroyl sarcosinate at pH 6.8 is presented in Figure 15. The SDS reaches a minimum surface tension of about 33 dynes cm^{-1} and the sarcosinate a minimum of 21.7 dynes cm^{-1}. A 20:80 mixture of sarcosinate:SDS reaches a minimum of about 24.5 dynes cm^{-1} and a 50:50 mixture, a value of 22.7 dynes cm^{-1}. When these mixtures reach a concentration approximately equal to the CMC of SDS, the surface tension begins to rise towards the value for pure SDS, possibly due to the solubilization of the sarcosinate by SDS micelles. However, at pH 4 the 50:50 mixture reaches a minimum surface tension of 24.5 dynes cm^{-1} and stays down. The CMC of the mixture is also depressed by a factor of about 5.

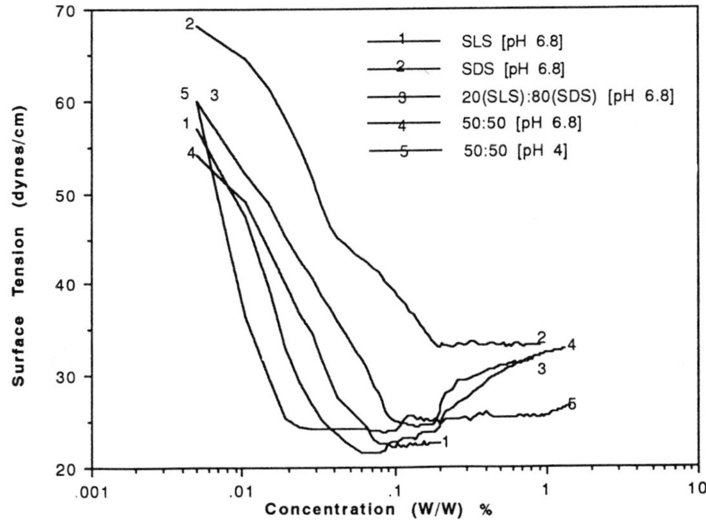

Figure 15 Sodium Lauroyl Sarcosinate and Sodium Dodecyl Sulphate

Applications:

Corrosion Inhibition:

The N-acyl sarcosines have been shown to be effective corrosion inhibitors in aqueous environments. The corrosion inhibition of steel alloy J55 in 15% HCl at 40°C by sarcosinates, is presented in Table 6. The weight loss of the specimens was taken as a measure of corrosion. All three sarcosinates reduce the rate of corrosion almost 10 fold over the blank. Under these severely acidic conditions however, the sarcosinates will slowly hydrolyze to sarcosine and fatty acid and lose their effectiveness.

Table 10

The Corrosion Inhibition of Mild Steel in 15% HCl at 40°C by Acyl Sarcosines

Acyl Sarcosines	%	% Weight Loss 24 Hours	% Weight Loss 96 Hours
Blank	0.0	3.75	20.37
L	0.5	0.53	3.5*
L	2.0	0.41	2.1
C	0.5	0.40	2.56*
C	1.0	0.35	2.23
C	2.0	0.68	2.09
O	0.5	0.45	4.5*
O	1.0	0.51	2.2
O	2.0	0.24	2.4

*Solution had changed color from yellow to black, possibly indicating that all the Acyl Sarcosine had broken down.

Pines and Spivack [9] subjected oleoyl sarcosine to a range of rusting tests designed to indicate its suitability for use as an oil soluble rust preventative in gasoline, lubricating oils and greases. They concluded that it acts as an effective rust inhibitor under dynamic and static type inhibition tests, as well as in the presence of corrosive atmospheres. They also found oleoyl sarcosine to be a very effective corrosion inhibitor in sea water and in silicone oils.

DeBerry and Viehbeck [10] showed that the addition of lauroyl sarcosine to an aggressive chloride solution decreases the susceptibility of type 304L stainless steel to corrosion and pitting.

Bartlett [11], in a patent assigned to DuPont de Nemours El & Co., discloses the use of sodium oleoyl sarcosinate as a corrosion inhibitor in tin plated aerosol cans.

Shell International [12] disclose the use of oleoyl sarcosine and morpholine as a corrosion inhibitor for city gas pipelines.

Oakes [13], in a patent assigned to Dow Chemical Company, teaches the use of sodium lauroyl sarcosinate with sodium nitrite to inhibit the corrosion of steel in an aqueous oxidizing environment.

Lubrication:

Murray and Westlund [14], in a patent assigned to Esso Research and Engineering, disclose the use of sodium nitrite and sodium lauroyl sarcosinate as a water based metal cutting and grinding fluid.

Calcium and barium salts of N-acylated sarcosines have been claimed as additives which markedly reduce soot in "smokeless"-type diesel engine fuels [15].

The alkali (particularly lithium) and alkaline earth salts of stearoyl sarcosine have been used [16] as lubricating grease thickeners and also offer the advantages of an antioxidant and corrosion inhibitor.

Herd and Horodysky [17], in a patent assigned to Mobil Oil Corporation, disclose the use of a partially borated alkyl diamine acyl sarcosine as an additive which confers exceptional friction reducing properties to lubricants.

Fuel Additives:

Nestor and Beck [18] using an electrochemical technique showed lauroyl sarcosine to be an effective corrosion inhibitor in wet isooctane. They showed that the sarcosine exerted both a hydrophobic and an electrochemical effect, presumably by polarization of local cell action. They concluded that the lauroyl sarcosine should be suitable as a corrosion inhibitor for aviation fuels.

Martin and Titsworth [19] showed that mixtures of oleoyl sarcosine with 1-hydroxyethyl 2-heptadecyl imidazoline or with N-alkyl propylene diamine are very effective corrosion inhibiting combinations for use in gasoline.

Fuels useful for rockets, missiles and projectiles can be produced by emulsifying hydrazine in mixed hydrocarbons with oleoyl sarcosine [20].

Eckert [21] found that aliphatic primary amine salts of N-acyl sarcosines were effective anti-stalling additives for gasoline.

Sung [22], in a patent assigned to Texaco, Inc., showed that by reacting N-acyl sarcosine with a poly alkylene polyol containing propylene oxide - ethylene oxide block copolymer units, he could produce a fuel additive that could reduce accumulation of deposits in the fuel intake, combustion chambers and exhaust system of internal combustion engines.

Leather Treatment:

It is desirable to render leather waterproof without inhibiting its breathability or air permeability. When leather is treated with a solution of sodium oleoyl sarcosinate [23] the solution permeates the leather. As it permeates, the pH of the solution drops due to the acidity of the leather. As the pH drops, the oleoyl sarcosine precipitates as an insoluble monolayer on the surface of the leather fibers, rendering them hydrophobic. The high contact angle for water on the coated surface of the fibers, prevents permeation. The adsorbed surfactant does not interfere with the permeation of air through the matrix.

Bay et al [24] in a patent assigned to BASF A. G., disclose the use of an emulsion of dimethyl siloxane in sodium oleoyl sarcosinate solution for waterproofing leather and hides. Testing of the treated leather in a Bally penetrometer revealed no breakthrough of water after 24 hours whereas the untreated leather failed after 4 minutes.

Carpet, Rug and Fabric Cleaning, and Static Reduction:

Mixtures of sodium lauroyl sarcosinate and sodium dodecyl sulfate have been widely used for in situ cleaning of carpets and rugs. The mixtures produce copious foam with strong detergency. The foam dries to a powdery non-tacky residue which can be easily vacuumed off. By contrast, sodium dodecyl sulfate alone produces a tacky residue which leads to rapid dirt redeposition. The presence of sarcosinate also leads to rapid static dissipation. Since the sarcosinate is a corrosion inhibitor it will protect the metal cans used for aerosol formulations, and other metal surfaces with which the shampoo makes contact.

Morgan et al[25], in a patent assigned to S. C. Johnson, teach the use of sarcosinates with methyl methacrylate-methacrylic acid copolymers in rug shampoo formulations to further reduce dirt redeposition.

Alpha alumina monohydrate, $Al_2O_3 \cdot H_2O$, which forms stable cationic dispersions in aqueous media, has been shown to produce more stable foams and drier residues, and further enhances static dissipation in sarcosinate containing formulations [26].

Wells et al [27] describe the production of an antistatic polyamide fiber by incorporation of a sarcosinate and a high molecular weight propylene oxide - ethylene oxide copolymer based on ethylene diamine. The sarcosinate alone was also shown to act as an antistat.

Fine Fabric Detergents:

The sarcosinates have been used as components of fine fabric detergents. They are mild enough to be suitable for hand use and have the ability to reduce the irritation caused by harsher components in the formulation. They minimize shrinkage and felting of woolens and dissipate static on synthetic fibers. Since they are compatible with cationic surfactants, these materials may be incorporated into the formulation to increase the degree of conditioning.

Bleach Formulation:

Sarcosinates have been widely used as components in thickened alkali metal hypochlorite bleach compositions [28, 29, 30]. Jeffrey and Grey, in a patent assigned to Jeyes Group Limited [31], disclose the use of sarcosinates in combination with quaternary ammonium compounds or amine oxides to produce thickened alkali metal hypochlorite bleaches which are stable with respect to viscosity and available chlorine content.

Sodium lauroyl sarcosinate, in combination with sodium dodecyl sulfate, has been used in an antimicrobial composition containing hydrogen peroxide at pH 4 to 6 [32]. The system was shown to be stable and noncorrosive to surgical instruments.

Froth Flotation:

Hellsten and Klingberg [33] in a patent assigned to Berol Kemi A.B., describe the synthesis and use of an amphoteric flotation agent synthesized from sarcosine and an alpha olefin oxide, which is useful for the froth flotation of calcium phosphate in the presence of calcium carbonate.

Calcium tungstate can be separated from quartz substrates using N-acyl sarcosines at pH 2.3 [34] and from stannic oxide by flotation with sodium cocoyl sarcosinate above pH 9 [35].

Baldauf et al [36] found oleoyl sarcosine to be an effective fluorite (calcium fluoride) collector in the presence of calcite (calcium carbonate), and sodium lauroyl sarcosinate has been shown to be an effective flotation agent for iron oxide fines [37].

Dispersants:

N-acyl sarcosines and their amine salts have been used as extrusion plasticisers for ceramics used in the production of electronic components and ceramic catalyst supports. When the ceramic base is fired, the surfactants burn away leaving no residue.

N-acyl sarcosines have been used as dispersants for magnetic media used in production of floppy disks, audio and videotapes. They are good dispersants, confer lubricity, have good dielectric properties and inhibit corrosion and oxidation.

Cruikshank et al [38] in a patent assigned to Eastman Kodak, have shown that a mixture of sodium dodecyl sulfate and sodium lauroyl sarcosinate in an aqueous hydrophilic colloid dispersion containing gelatine and silver halide, provides excellent properties with regard to uniform spreading and reduction of repellency defects when the dispersion is applied as a coating on photographic substrates.

Vlay et al [39], in a patent assigned to Carl Zeiss Jena, showed that when oleoyl sarcosine was added to conventional lapping agents and fluids, more material was removed and products of improved quality were produced.

Biotechnology:

Sodium lauroyl sarcosinate has been shown by Filip et al[40], to selectively solubilize the cytoplasmic membrane of lysed Escherichia Coli cells, while leaving the outer lipopolysaccharide membrane intact under conditions in which Triton X-100 and sodium dodecyl sulfate solubilized all membrane protein. The presence of Mg^{2+} during treatment with sarcosinate afforded partial protection of the cytoplasmic membrane from dissolution.

Tremblay et al[41] developed a technique for detecting deoxyribonucleic acid (DNA) associated with membrane. In this procedure, cells lysed in the presence of sarcosinate and Mg^{2+} produced a consistent band on centrifugation which contains crystals of Mg^{2+} sarcosinate, membrane and attached DNA.

The hormone, bovine somotrotropin, has great commercial potential, as a growth rate promoter in cattle, hogs and salmon which also increases the meat to fat ratio. It also increases milk production in dairy cattle. Because of their capability to selectively solubilize proteins without causing them to be denatured, sarcosinates are used to extract bovine somatotropin from transformed E coli cells [42].

Pesticide Formulation:

Aldwinkle[43], in a patent assigned to Ciba Geigy, showed that sodium sarcosinates functioned as compatibilizers in pesticide - liquid fertilizer formulations. He showed that the stability of a mixture of isoproturon (a herbicide for control of grasses and broad leaved weeds) and a liquid fertilizer could be enhanced by addition of sodium lauroyl sarcosinate.

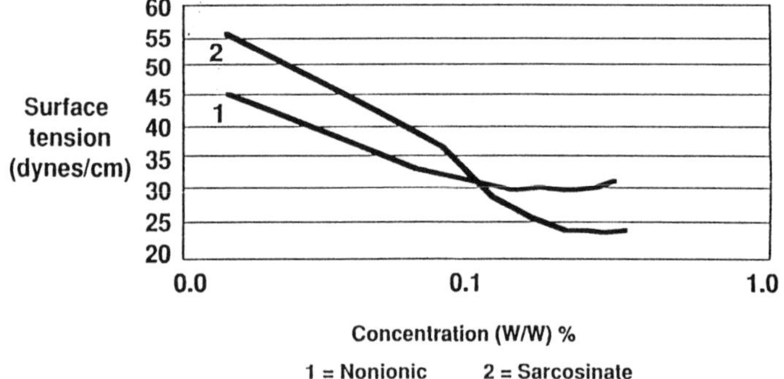

Figure 16 **Glyphosate Pesticide with**
1) Ethoxylated Nonionic Surfactant
2) Sodium Cocoyl Sarcosinate

Sarcosinates may be used as surfactants in herbicides based on isopropylamine salts of glyphosate. Nonionic tallow ethoxylates have been used conventionally in this type of herbicide. When sodium cocoyl sarcosinate is used as a surfactant, as can be seen from Figure 16, a minimum surface tension of about 25 dynes cm^{-1} is reached whereas with the nonionic surfactant the minimum surface tension possible is 30 dynes cm^{-1}. This depression of the surface tension by an extra 5 dynes cm^{-1} could potentially allow the herbicide to penetrate the plant through the stomata[(44)].

Plastics:

Sarcosinates have found application in many areas of the plastics industry.

Mold Release:

It is necessary in "Reactive Injection Molding", RIM, to provide adequate release from the mold in the shortest possible time. Mayer et al [(45)], in a patent assigned to Dow Chemical, teach the use of oleoyl sarcosine as an in situ mold release agent for reactive injection molding of polyether polyurethanes. The sodium oleoyl sarcosine is apparently squeezed from the lattice as the resin crosslinks and acts as a release agent by occupying the interface between mold and resin.

Non-Fogging Plastics:

Hammond [(46)], in a patent assigned to National Distillers and Chemical Company, details the production of non-fogging thermoplastic polyolefin sheets, by incorporation of 0.001% of an N-acyl sarcosine. The sheets also exhibited lower haze, higher gloss, better adhesion of ink, coatings and labels and increased oxygen permeability which aids in keeping meat looking fresh.

Antistatic Plastics:

N-acyl sarcosines are F.D.A. approved as antistatic and anti-fogging agents at up to 0.15% by weight in polyolefin films for packaging meat, fruit and vegetables[(47)].

Cure Time Extender:

The cure time of a moisture cured polyurethane prepolymer may be prolonged by the addition of sodium lauroyl sarcosinate[(48)].

Emulsion Polymerization:

Sarcosinates have found use in emulsion polymerization because they are effective emulsifiers, give good rates of polymerization, high percentage conversion and produce latices of small particle size. They also aid in reducing reactor fouling thereby simplifying reactor cleaning. Furthermore, as corrosion inhibitors, they can confer corrosion inhibiting properties to the finished resin. Because the surfactant is insoluble under acid conditions the latex may be coagulated by addition of acid, allowing the aqueous phase to be drawn off.

Personal Care:

Dentifrice Applications:

Tooth decay is caused by the action of bacteria on sugars producing acid in the mouth.

In the 1950's, King [(49)], in a patent assigned to Colgate-Palmolive showed that sodium lauroyl sarcosinate, Gardol™*, when incorporated into dentifrices had the capability to inhibit dental caries for more than 24 hours, after a single washing.

The sarcosinate is believed to act as a hexokinase inhibitor and it's prolonged effectiveness was attributed to its strong adsorption on dental plaque.

Tomlinson [(50)] has shown by an in vivo experiment, that a mouthwash containing 0.5% sodium lauroyl sarcosinate, when used 3 times a day, completely stopped the loss of dentine from the test specimen, even in the presence of high levels of sucrose in the mouth.

The fact that sarcosinates are good foaming agents and detergents and have a mild taste has allowed them to be used as the sole surfactant in many dentifrices.

Shampoos and Liquid Soaps:

Sarcosinates have found widespread use in shampoos and liquid soaps [(51, 52, 53)].

The advantages which sarcosinates confer to these products include effective foaming and detergency below pH 7, and substantiveness to hair and skin which leads to conditioning, easy compatibility and dissipation of static. The compatibility of sarcosinates with cationic surfactants allows varying degrees of conditioning to be achieved. The mildness of sarcosinates along with their ability to reduce the irritability of other surfactants is also important.

Syndet Soap Bars:

It is possible to produce synthetic detergent soap bases by using warm N-acyl sarcosine as a solvent for fatty acids such as stearic acid or lauric acid and neutralizing the mixture to pH 6 - 7 with a base such as sodium hydroxide [(54)]. The mixture which is a clear liquid at 40 - 50°C, solidifies rapidly on cooling to a hard soap like material which produces copious foam and an elegant skin feel. The material can be molded into soap bars, but since it can be remelted it can also be processed by conventional techniques.

* Colgate-Palmolive

References:

1) W. C. Griffin, J. Cosmetic Chem., 1, 311 (1949); ibid., 5, 249 (1954).

2) J. T. Davies, Proc. 2nd Int. Congr. Surf. Act., London, Vol 1, p 426.

3) M. F. Nelson, Jr. and D. Stewart, Jr., Society of Cosmetic Chemists, Sept. 15-16, 1955, New York City.

4) G. A. Salensky, M. G. Cobb and D. S. Everhart, Ind. Eng. Chem. Prod. Res. Dev., 1986, 25, 133-140.

5) F. R. Bettley, Brit. J. Dermatol., 77, 98-100 (1965)

6) V. Dvorkovitz, U. S. Patent 2,962,448, assigned to Diversey Corporation.

7) J. R. Hart and M. T. DeGeorge, J. Soc. Cosmet. Chem., 31, 223-236 (Sept/Oct 1980)

8) M. J. Rosen, Surfactants and Interfactial Phenomena, 2nd Ed., John Wiley & Sons, New York, 1989.

9) R. M. Pines and J. D. Spivack, Corrosion, 13, 690t-694t (1957).

10) D. W. DeBerry and A. Viehbeck, J. Electrochem. Soc., p30-36, Jan 1986.

11) P. L. Bartlett, U. S. Patent 4,602,958, assigned to DuPont deNemours & Co.

12) Shell Internationale Research, Maatschappij. N. V., Netherlands Appl. 6.607,803.

13) B. D. Oakes, U. S. Patent 2,931,700, assigned to Dow Chemical Co.

14) J. L. Murray and R. A. Westlund, U. S. Patent 3,046,225 (1962).

15) BASF A-G, British Patent 1,267,275 (March 1972).

16) J. F. Lyons and P. R. Thomas, U. S. Patent 2,841,555, assigned to Texas Co. (1958).

17) R. S. Herd and A. G. Horodysky, U. S. Patent 4,474,671.

18) L. J. Nestor and W. Beck, Materials Protection and Performance, Vol. 9, No. 12, Dec. 1970.

19) E. C. Martin and H. R. Titsworth, U. S. Patent 2,919,979, assigned to American Oil Company.

20) B. A. H. Seaman, U. S. Patent 3,743,555, assigned to Dow Chemical.

21) G. W. Eckert, U. S. Patent 3,109,719, assigned to Texaco, Inc. (1961).

22) R. L. D. Sung, U. S. Patent 4,758,247, assigned to Texaco (1988).

23) R. J. Holzinger and G. M. Whelan, U. S. Patent 2,964,425, assigned to Socony Mobil Oll Company.

24) H. Bay, R. Widder and G. Eckhert, Ger. Offen. DE 3,529,869, assigned to BASF A-G.

25) L. W. Morgan, N. D. Shrikishna, U. S. Patent 3,723,323, assigned to S. C. Johnson & Son, Inc.

26) L. J. Schuman and P. J. Carlucci, Soap & Chemical Specialties, p43-70, March 1970.

27) R. L. Wells and L. Crescentini, U. S. Patent 3,808,291.

28) U. Schilb, U. S. Patent 4,337,163, assigned to Lever Brothers Co. (1982).

29) H. L. Diamond and T. L. Murphy, U. S. Patent 4,388,204, assigned to The Drackett Co. (1983).

30) R. J. Cramer and W. L. Smith, European Patent Appl. 87,305,455.5, applied for by The Clorox Company, (1987).

31) D. J. Jeffrey and R. A. Grey, British Patent 1,466,560, assigned to Jeyes Group Ltd.

32) G. W. Willcockson and D. C. F. Law, European Patent Appl. 83,306,894.3, applied for by Surgikos, Inc.

33) K. M. E. Hellsten and A. W. Klingberg, U. S. Patent 4,358,368, assigned to Berol Kemi AB.

34) R. Von Gathlen, Bergauwissen Schaften, 7, 352, 1960.

35) H. Schranz and R. VonGathlen, Bergauwissen Schaften, 7, 360, 1960.

36) H. Baldauf, H. Schubert and W. Kramer, Aufbereit-Tech, 27 (5), 235-41, 1986.

37) P. Young, Z Erzbergbau Mettalhutten W., 14 (4), 184-188.

38) S. J. Cruikshank and M. W. Orem, U. S. Patent 4,370,412, assigned to Eastman Kodak Company, (1983).

39) K. H. Vlay, D. Landgraf, V. Wendel, R. Fleischauer, and J. Fuchs; Ger. (East) Patent DD233,135, assigned to Carl Zeiss Jena.

40) C. Filip, G. Fletcher, J. L. Wulff and C. F. Earhart; Journal of Bacteriology, p717 -722, Sept 1973

41) G. Y. Tremblay, M. J. Daniels and M. Schaechter; J. Mol. Biol., 40, p65-76, 1969.

42) T. Evans, International Patent Appl. PCT/US86/01289, applied for by the Upjohn Company.

43) M. J. Aldwinckle, Offen DE 3,308,187, assigned to Ciba Geigy A-G (1983).

44) P. J. Holloway and S. Stock, "Factors Affecting the Activation and Foliar Uptake of Agrochemicals", Industrial Applications of Surfactants II, Edited by D. R. Karsa, Royal Society of Chemistry, 1990.

45) L. W. Meyer, J. A. Vanderhider and R. Carswell, U. S. Patent 4,876,019, assigned to Dow Chemical Co.

46) J. E. Hammond, Canadian Patent 844,673, assigned to National Distillers and Chemical Corporation.

47) Federal Register, Vol 31, No. 248, Friday, December 23, 1966.

48) A. J. Gasper and J. N. Wells, European Patent Appl. 82,301,884.1, applied for by Minnesota Mining and Manufacturing Co.

49) W. J. King, U. S. Patent 2,689,170, assigned to Colgate-Palmolive Company, (1954).

50) K. Tomlinson, J. Cosmet. Chem., 29, p385-397, 1978.

51) J. R. Hart, "Sarcosinate Surfactants in Skin Cleansers", Cosmetic Technology, Jan 1980.

52) J. R. Hart, "Liquid Soap - A Challenge for the Formulator", Household and Personal Products Industry, May 1981

53) J. R. Hart and E. F. Levy, "Sarcosinate-Cationic Creme Rinse Shampoos", Soap Cosmetics and Chemical Specialties, August 1977.

54) Patents Pending, J. J. Crudden, W. R. Grace & Co.-Conn.

Nonionic Surfactants

Alkylglycoside Surfactants — Synthesis and Properties

Joachim Thiem and Thomas Böcker

INSTITUT FÜR ORGANISCHE CHEMIE, MARTIN-LUTHER-KING-PLATZ 6, D-2000 HAMBURG 13, GERMANY

Introduction

Annually by photosynthesis approximately 170 billion tons of biomass are produced.[1] Via the central precursor glucose,other monosaccharides, oligosaccharides, and polysaccharides are formed. A series of well established metabolic pathways lead from glucose to proteins, alkaloids, porphyrins, nucleic acids, fatty acids, terpenes and steroids, phenols, lignin and aromatic amino acids, hence to all the other compounds of the life cycle.[2] Carbohydrates and derivatives comprise some 95 % of the total with the supporting material cellulose dominating by far. Presently less than 3 % of this amount (still more than 4 billion tons/a) is used by men predominantly for feeding purposes as evident from the table of distribution (cf. Figure 1).[1]

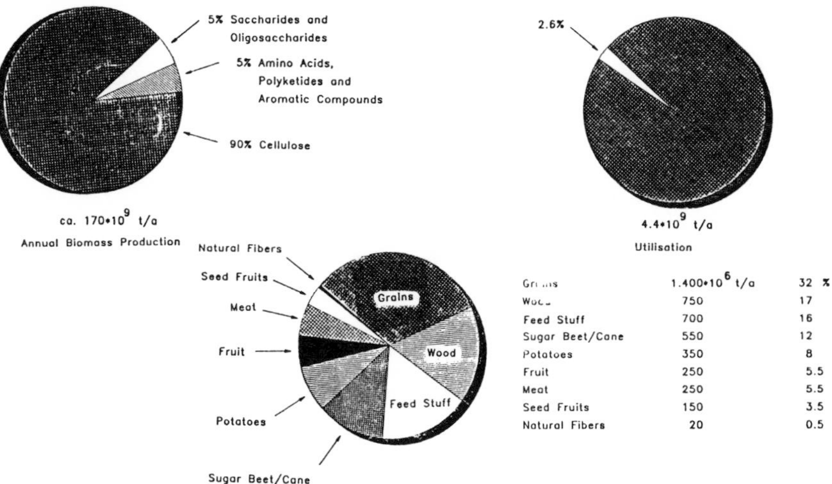

Gr[illegible]s	$1.400 \cdot 10^6$ t/a	32 %
Woo[illegible]	750	17
Feed Stuff	700	16
Sugar Beet/Cane	550	12
Potatoes	350	8
Fruit	250	5.5
Meat	250	5.5
Seed Fruits	150	3.5
Natural Fibers	20	0.5

Figure 1 Biomass and Utilisation

In the long run it may be necessary and presently it is in vogue and interesting to imagine a more intensive use of biomass by men with respect to production of other goods from regrowing material. Certainly their properties will have to be compared and must compete in all respects with existing materials. However, in addition to classical measures for comparision increasingly answers concerning the degradation properties will be required. Thus, an integrated faster recycling of any compound or material into the metabolic processes may be an attractive goal for development. As depicted in Fig. 2[3] a carbohydrate-based chemical industry[4] may be presently a vision with respect to a complete satisfaction of all demands. However, a solid part of compounds emanating from the indicated industrial section are even nowadays based on regrowing resources, e.g. approximately 10 % of the raw material used in the German chemical industry is based on natural compounds.[5]

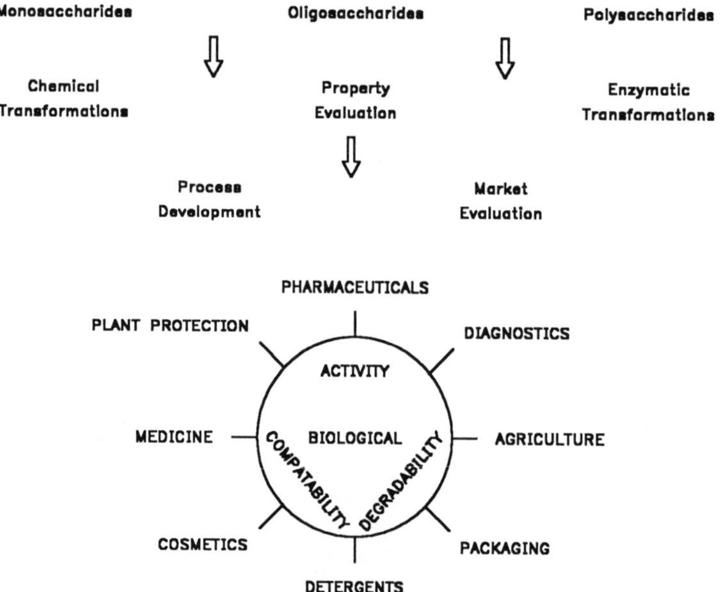

Figure 2 Carbohydrate-Based Chemical Industry

The use of saccharides especially for surfactants represents a particularly attractive approach, and first ideas and realisation in this direction[6] are certainly based on natural precursors. Several glycolipids such as rhamnose (**1**), sophorose (**2**), and trehalose lipids (**3**) (Fig. 3) were 'invented' early on by aqueous-system based microorganisms who feed on aliphatic or aromatic carbon sources virtually insoluble in water.[7,8] Simply by conjugation of hydrophilic sugars with hydrophobic fatty acids they form these biosurfactants which nicely serve to metabolize hydrocarbons.

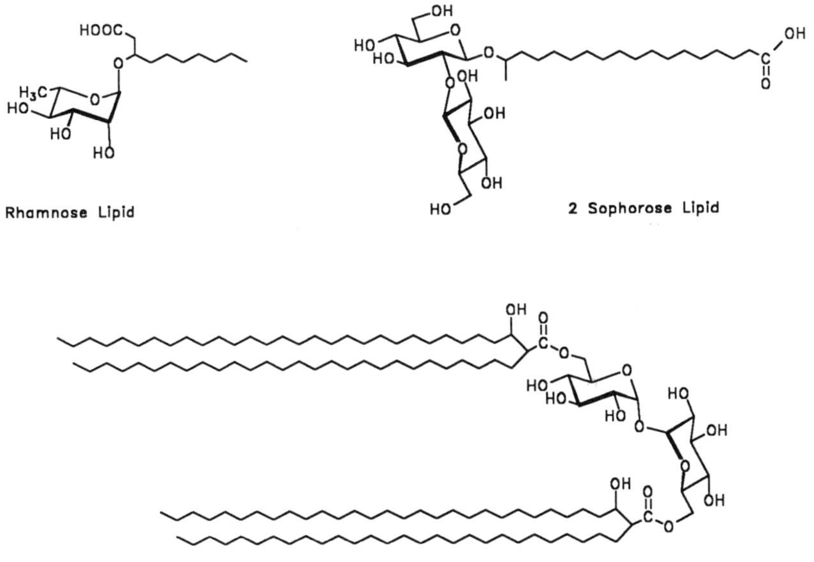

1 Rhamnose Lipid

2 Sophorose Lipid

3 Trehalose Lipid

Figure 3 Biosurfactants

Previously studies on the metabolic degradation of radioactively labelled dodecyl ß-maltoside (**4**) were performed.[9,10] These revealed the assumed cleavage of the interglycosidic bonds to give glucose and dodecyl ß-glucoside (**5**). The latter was further cleaved directly to glucose and the fatty alcohol, or transformed into the 6-O acyl derivative **6** and then cleaved completely. The resulting sugar will be incorporated into the carbohydrate metabolism and the fatty alcohol eventually into the fatty acid metabolism. (cf. Fig. 4)

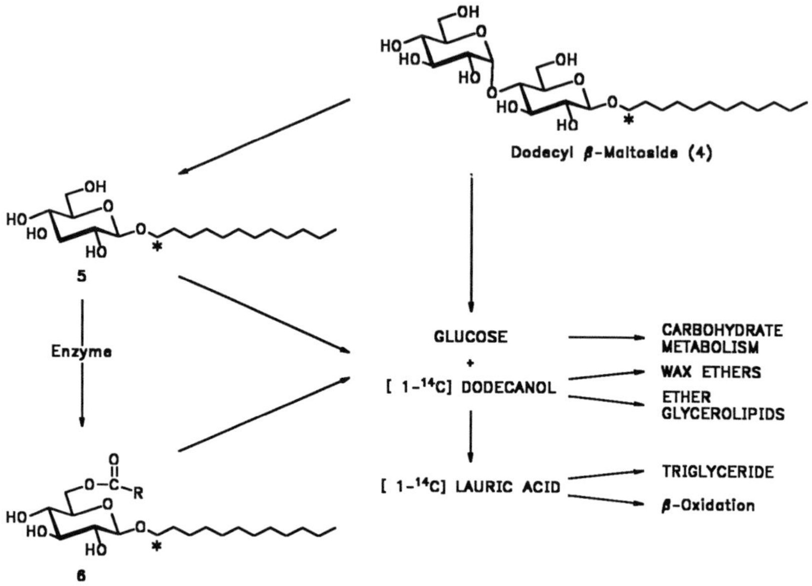

Figure 4 Metabolic Degradation of Alkyl Glycosides

A number of years ago and even now and then today studies of sugar-fatty acid esters report on attractive properties of these derivatives as emulsifiers particularly in the food area. Predominantly sucrose esters were studied closely.[11] However, for a wider range of applications esters proved to be too labile, and thus the interest focused on the formation of more stabile acetals.

Based on earlier studies industrial laboratories reported on the formation of alkyl polyglucosides (APG) following the Fischer glycosylation method.[12] Owing to the low solubility of glucose in a long chain fatty alcohol the common approach uses a two stages process. First, glucose (**7**) is transferred with butanol in the presence of a proton source such as p-toluene sulfonic acid predominantly into butyl α-D-glucopyranoside (**8**). As known from several model studies a number of other butyl glycosides having various ring sizes, anomeric linkages and chain lengths are also formed. This precursor mixture is fairly well soluble in a fatty alcohol mixture e.g. obtained from coconut oil and thus a subsequent Fischer glycosylation gives the products **9** again together with other material (Fig. 5). Even though several studies on the exact structure of APG were started, still the type of linkages, the regiochemistry, the degree of polymerization and the like are difficult to assess.

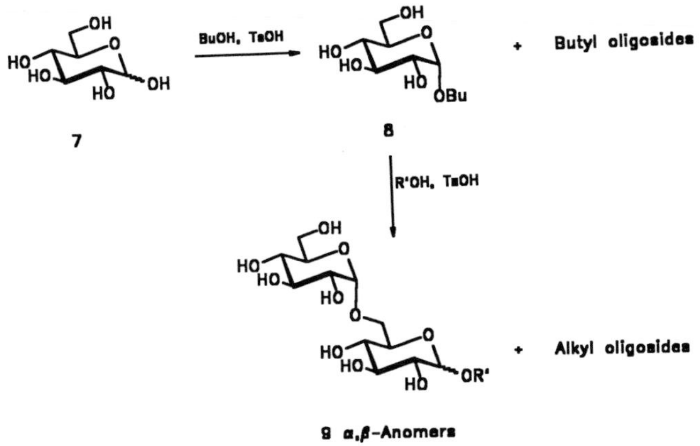

Figure 5 Formation of APG by a Two Stage Fischer Glycosylation

Our interest focused on the systematic formation of a series of alkyl glycosides in order to obtain closer information on structurally dependent properties of the individual components that make up commercial APG mixtures. Therefore, glucose and some other hexoses were transformed into various anomeric alkyl glycosides to prove the influence of configuration, anomeric structure and alkyl chain length on surfactant properties. Further, a series of alkyl glycosides of α, 1 - 4 glucanes (maltooligosaccharides) and also of the α, 1 - 6 glucane type (isomaltooligosaccharides) were prepared and the change of properties studied.

Glycosylation

Koenigs-Knorr-Type Reactions

Even though in Fischer glycosylations[12] α-glycosides prevail owing to the anomeric effect and thermodynamic control the usual approaches for selective formation of glycosides require considerably more sophisticated methods.[13] Most frequently applied are variations based on the Koenigs-Knorr method[14] which make use of a peracylated glycosyl halide (chloride or bromide) in turn obtained from the peracylated sugar derivative. In the presence of promoters such a silver or mercury salts nucleophilic attack will lead to the glycosylated sugar peracylates, which are to be deesterified in the final step. Altogether, starting with the free sugar this approach requires four steps, costly or toxic promoter salts, and often additional separation efforts.

Previously, some model alkyl glycosides of the desired type were prepared following this approach.[15-19] An internal more favourable arrangement of the glycosyl halide on the surface of the solid catalyst is assumed to give rise to a S_N2 type attack of the nucleophile, and thus almost exclusive formation of 1,2-trans glycosides.[20] The yields observed for glycosylation of glucose, galactose, and mannose with octanol, decanol, and dodecanol range between 40 and 50 % over all steps.[18,19] Often a column separation is required which is favourably performed using strongly basic ion exchange resins (Dowex 1 x 2, OH^-).[21,22]

Obviously, it is highly desirable to employ a simpler and less costly approach to glycosides. Further, a major drawback of the Koenigs-Knorr type processes is the exclusive formation of the 1,2-trans products. However, since industrial APG synthesis by Fischer glycosylation predominantly leads to the 1,2-cis glycosides a simple yet stereochemically defined approach to α-glycosides of maltooligomers for comparison of analytical and physicochemical data is required.

Lewis Acid-Catalysed Glycosylation

An advantageous glycosylation procedure based on earlier studies of Zemplen,[23] Lindberg,[24] Lemieux[25] and Bundle et al.[26] with stannic tetrachloride as Lewis acid catalyst and ß-peracetylated sugar derivatives could be successfully employed in the synthesis of models for APG. As exemplified with the synthetic scheme (Fig. 6) treatment of maltotriose (**10**) with acetic anhydride and molten anhydrous sodium acetate at 120 °C gives exclusively the ß-undecaacetate **11**. For activation with the Lewis acid exclusively the ß-acetate is required, and reaction in dichloromethane with lauryl alcohol and stannic tetrachloride leads to the peracetylated lauryl maltotriosides **12α** and **12ß**.

Under these conditions the leaving group properties of the ß-anomeric acetate group is sufficient and the remaining oxocarbonium intermediate is stabilised by the neighbouring 2-acetoxy group to give a 1,2-acetoxonium intermediate. This in turn will be opened trans to give the ß-lauryl glycoside **12ß**. This ß-product of kinetic control forms at lower temperature and short reaction times. In contrast, at extended reaction times and at higher temperatures an in-situ transformation into the thermodynamically favoured α-glycoside **12α** is observed.

Following chromatographic purification the peracetylated products are deesterified with sodium methoxide and the anomers separated on strongly basic ion exchange resins to give dodecyl α-maltotrioside (**13α**) or dodecyl ß-maltotrioside (**13ß**).

In this example the decisive glycosylation step was performed at room temperature for 48 hours, and the yield of **13α** over four steps including separations was 49 %. Similarly, treatment of **11** with docecanol/$SnCl_4$ at 0-5 °C for 2 hours gave **13ß** in 36 % overall yield. Another advantage of the stannic chloride glycosylation is that only a 1.5 molar equivalent of alcohol is required, in contrast to Koenigs-Knorr approaches which often need a 10-fold excess of alcohol.[17,27]

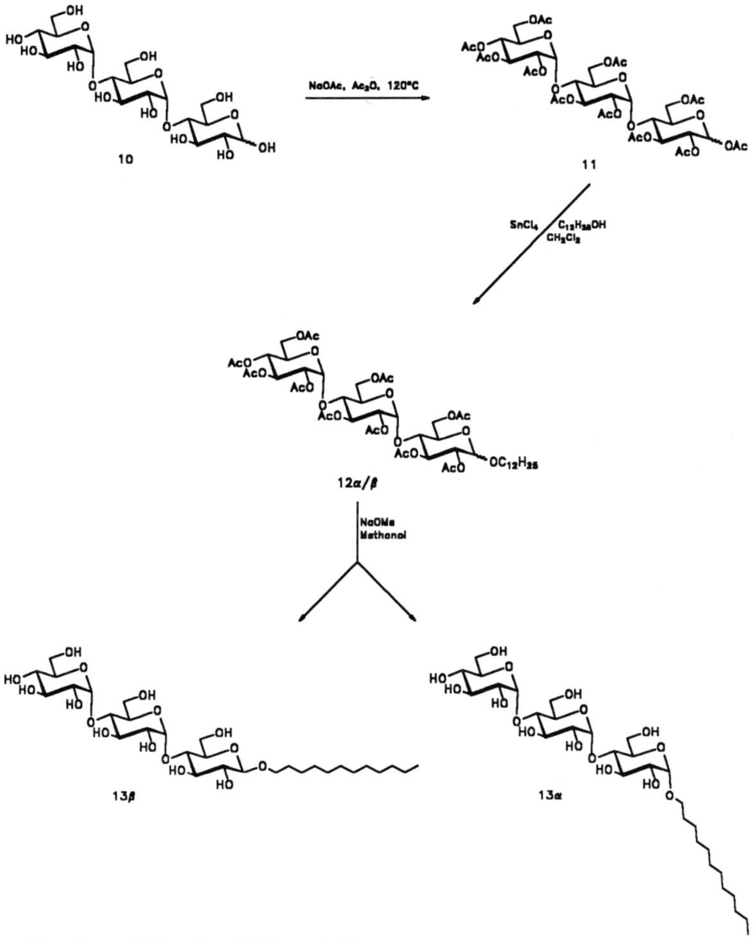

<u>Figure 6</u> Lewis Acid Catalyzed Glycosylation

For structural assignments extensively n.m.r. studies were employed. The example shown in Fig. 7 depicts the 2D-COSY representation of dodecyl 2,3,4,2',3',4',2",3",4",6"-deca-O-acetyl-ß-isomaltotrioside (**14**) which allows a straightforward interpretation and structural verification. Often peracylated derivatives are preferred for ^{1}H n.m.r. studies, and for the check of assignments useful literature compilations are available.[28-33]

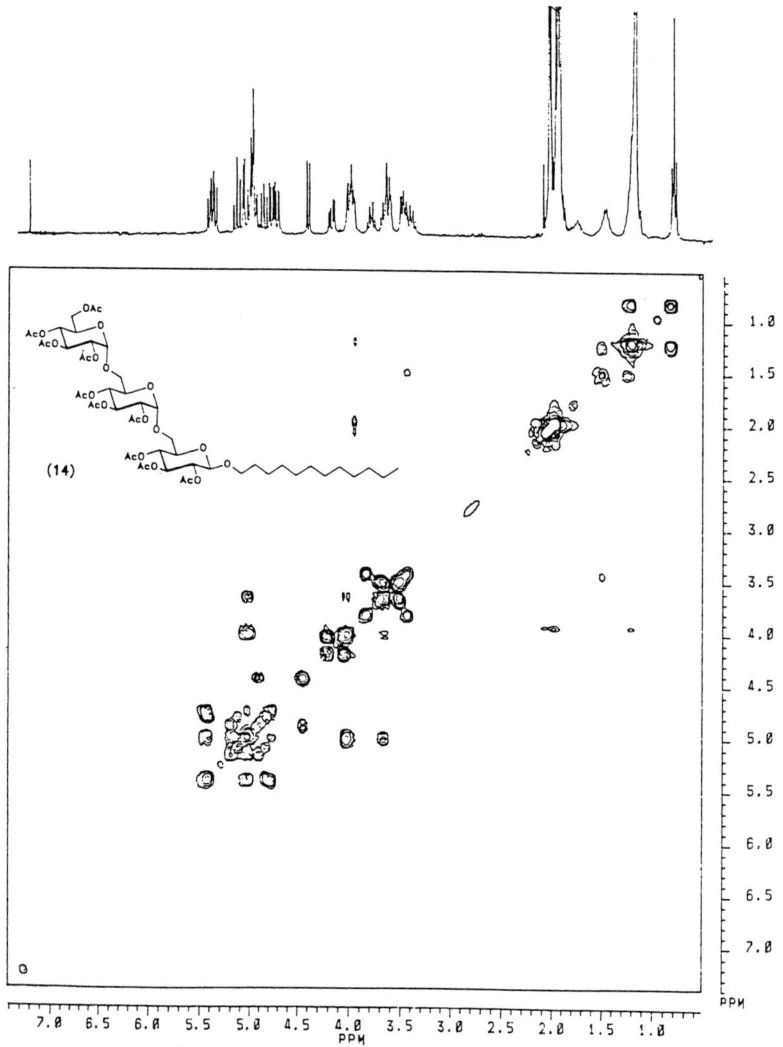

*Figure 7 2 D-COSY ^{1}H-NMR of Dodecyl β-Isomaltotrioside (**14**)*

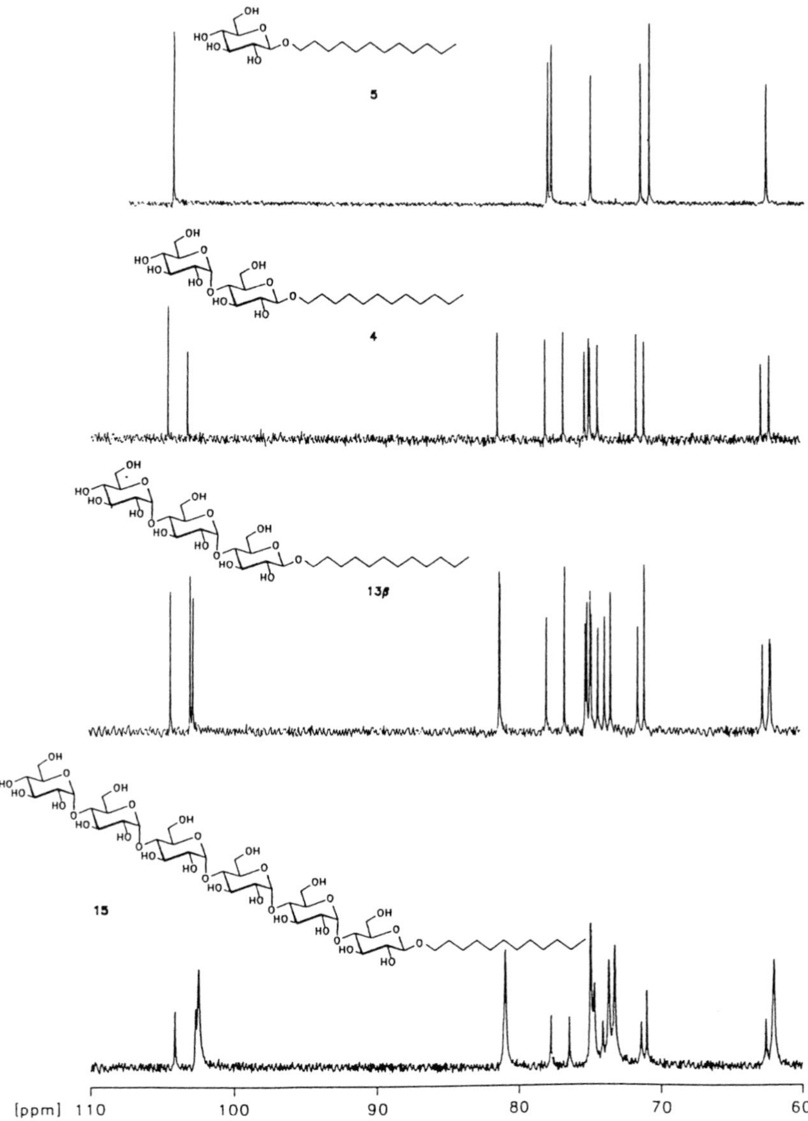

Figure 8 ^{13}*C-NMR-Spectra of Dodecyl β-Oligosaccharides*

The method of choice for the structure elucidation of unblocked higher oligosaccharide glycosides proves to be ^{13}C n.m.r. Up to the trisaccharide each ring carbon atom can be assigned. The anomeric carbons appear in the region of δ 105 (ß-interglycosidic linkage) or δ 100 (α-linked), C-4 at about δ 80 and C-6 at 66-69. The remaining ring carbons and the OCH_2 group show at δ 69-78 and then C-6 at 62 and alkyl chain carbon atoms at δ 14 and 33. In Fig. 8 the ^{13}C n.m.r. spectra of dodecyl ß-glycosides of glucose (**5**), maltose (**4**), maltotriose (**13ß**), and maltohexaose (**15**) are depicted. Obviously the assignment is easy up to the trimer, beyond that broader yet interpretable bands are observed.

Ferrier Glycosylation

An attractive approach to modified glycosides in using even weakly nucleophilic agents was developed by Ferrier et al.[34-36] As depicted in Fig. 9 ß-maltose octaacetate (**16**) on treatment with hydrogen bromide/acetic acid leads to the glycosyl bromide **17** and reductive elimination gives the peracetylated maltal **18**.[37-38] Glycosylation with dodecanol and borontrifluoride etherate proceeds by a nucleophilic attack at the anomeric center of the glycal. By an allylic rearrangement the activated ester function at C-3 leaves the molecule to give the 2,3-unsaturated glycoside **19α/ß** with the α-anomer largely prevailing (**19α** : **19ß** = 86 : 14). Hydrogenation on palladium/charcoal proceeded quantitavely, and on this stage separation of the anomers was performed by preparative HPLC. Following treatment with methoxide both the α-anomer **21α** and the ß-anomer **21ß** were obtained easily. In comparision to dodecyl α- and ß-maltosides (**31** and **4**) which are nicely soluble in water at room temperature both the 2,3-dideoxy analogues show very little solubility under these conditions. Further, a strong influence of the 2- and 3-hydroxy groups on the pKa value is evident from the rather different elution pattern on strongly basic ion exchange resin. The dideoxy glycosides **21α** and **21ß** are eluted early on, even before dodecyl α-(**51**)[39] and ß-glucosides (**5**), and only much later the parent components dodecyl α- and ß-maltosides (**31** and **4**) are eluted.

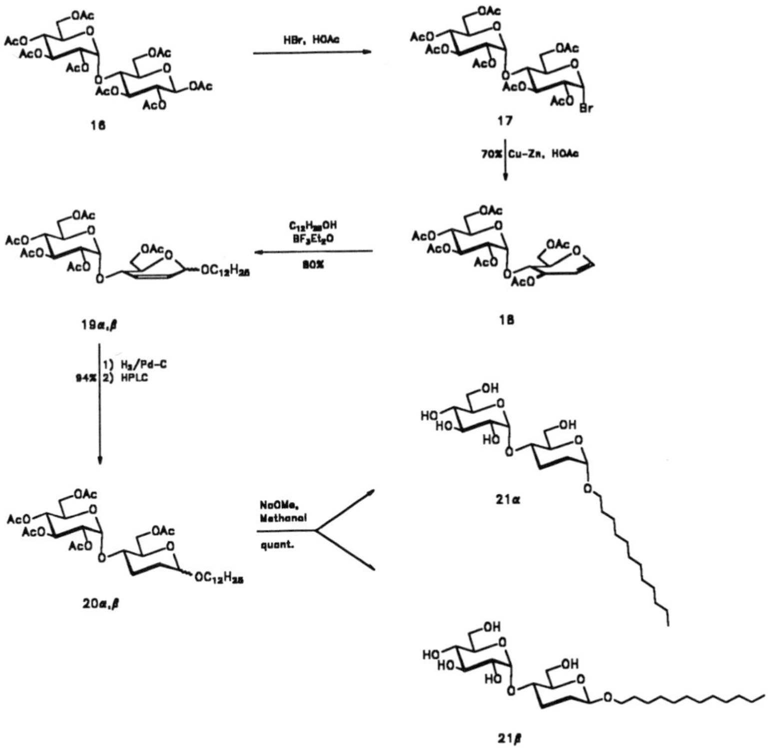

Figure 9 Modified Alkyl Glycosides by Ferrier Glycosylation

Physicochemical Properties

Measurements of the surface tension and the critical micelle concentration (c.m.c.) employed the ring method.[40]

The hydrophobic group has the most pronounced effect on the c.m.c. In the monosaccharide series octyl, decyl, and dodecyl glycosides of glucose (**22, 23, 5**), of mannose (**24 - 26**), and of galactose (**27 - 29**) show drastically decreasing c.m.c. with increasing alkyl chain length (Fig. 10, logarithmic scale). A similar observation applies to dodecyl and tetradecyl ß-maltoside (**4** and **30**) as well.

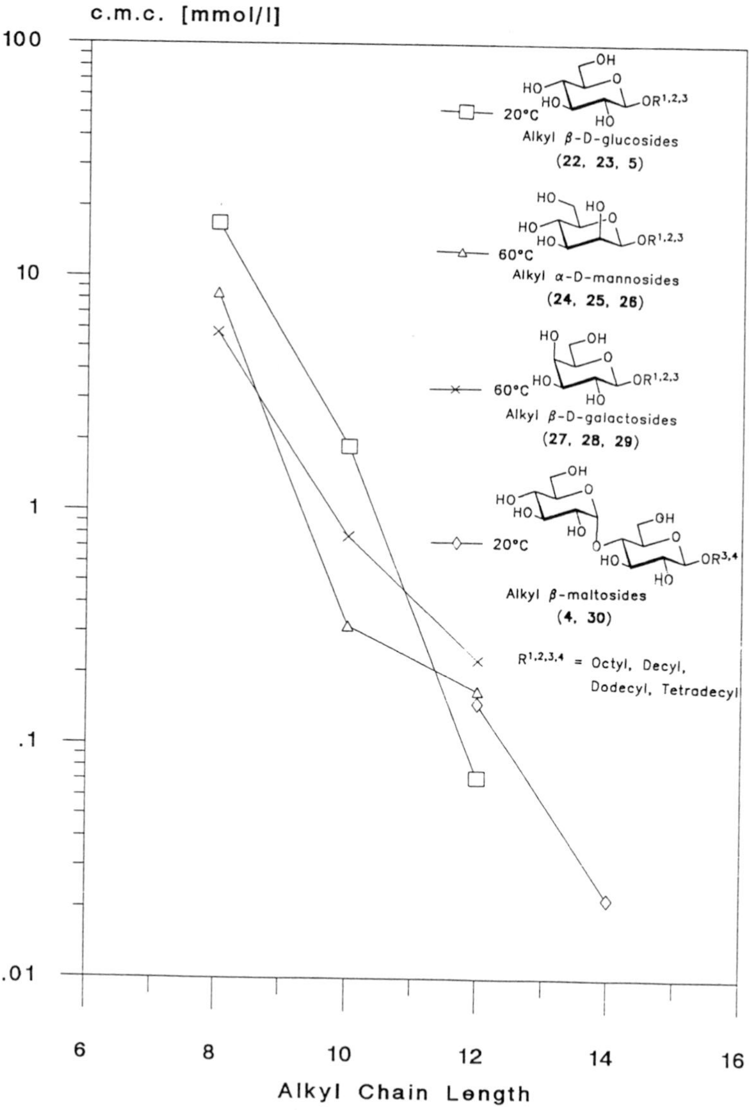

Figure 10 Correlation of CMC and Hydrophobic Group

In contrast, the effect of the hydrophilic part of the molecule seems to be of minor influence. Obviously, as depicted in Fig. 11 (normal scale), there is little difference in c.m.c. in going from dodecyl glucoside (**5**) to maltosides (**31** and **4**) and further to maltotriosides (**13α** and **13ß**). It may be noted that there is hardly more than a very small difference in c.m.c. between the α- and ß-anomers (the dodecyl α-glucoside is not soluble at 20 °C).

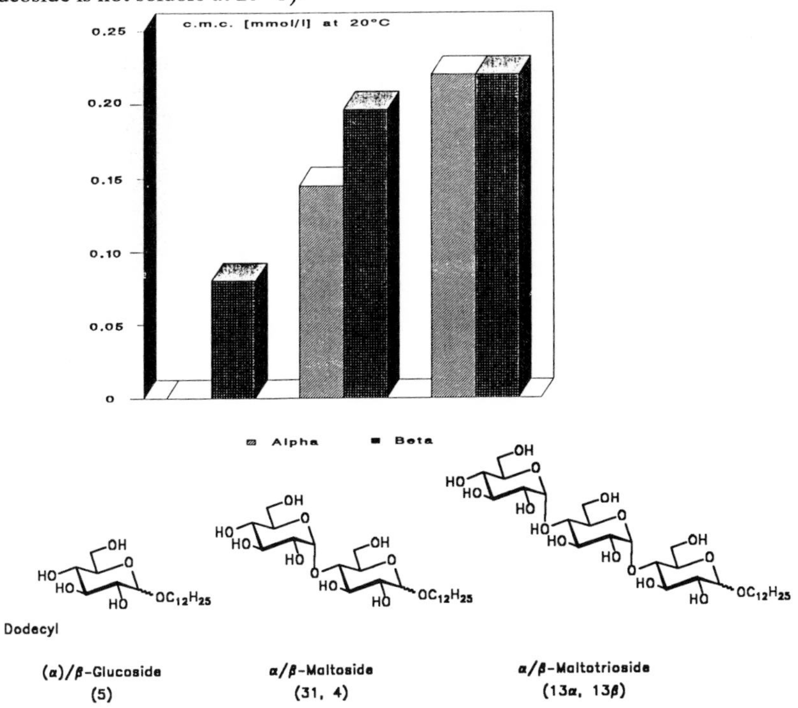

Figure 11 Correlation of CMC and Hydrophilic Group and Anomerity

In the series of dodecyl ß-maltooligomers (Fig. 12, left; compounds **5**, **4**, **13ß**, and **32-35**) the extension of the saccharide unit from one sugar to seven units gives rise to an increase in c.m.c. only by approx. a factor 4. It is surprising to note that even the heptaoside **35** shows the features of a nonionic surfactant, and it may be assumed that this correlates with a remaining helical structure based on intramolecular hydrogen bridges.

As previously discussed[19,41] for the strong influence of the hydrophobic group on c.m.c. another striking example is shown in Fig. 12 (right) for dodecyl and tetradecyl α- as well as ß-maltosides (**31, 4, 36,** and **30**). Here an increase in carbon chain length by one CH_2-group results in a decrease of c.m.c. by the factor 7 and 13, respectively.

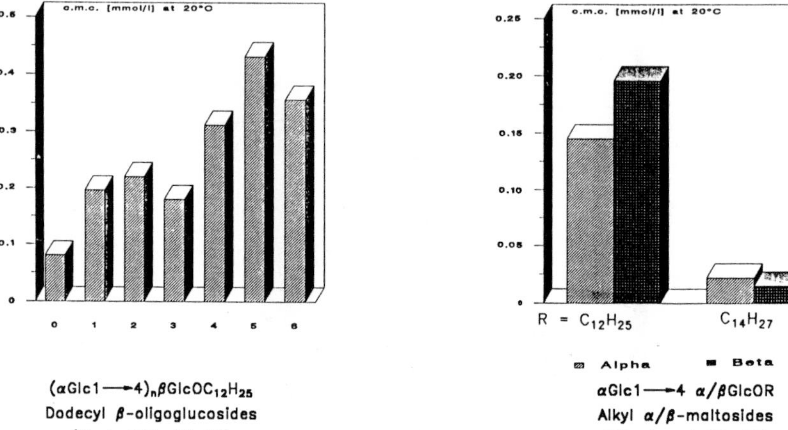

Figure 12 Influence of Hydrophilic and Hydrophobic Structural Elements on CMC

Analysis of APG Samples

Some analytical studies of selectively prepared APG samples were performed in order to elucidate partial structures of dominant components. Instead of a fatty alcohol mixture dodecanol was used throughout. It was of interest to gain some insight into the molecular weight distribution as well as amount and type of the dominating interglycosidic linkages. This requires a separation according to molecular weight first. Owing to their amphiphilic structure in aqueous solution only a small amount of long chain alkyl glycosides will be present as individual monomers, however, micell formation will occur. Therefore attempts to apply gelchromatographic separations in water were unsuccessful. Subsequently all samples were peracetylated and then separation could be performed by HPLC. In Fig. 13 two different APG samples and the butyl oligoglucoside mixture are shown together with a standard calibration based on a series of prepared oligomeric models in peracetylated form.

The so called 'hydrophobic APG, 2:7' was prepared from butyl glucoside and dodecanol in the ration 2:7 with p-toluene sulfonic acid as catalyst at 120 °C. Studies of the unblocked mixture of compounds was impossible due to micelle formation. Thus, the reaction mixture was peracetylated and then separated by gelpermeation HPLC on Shodex HF 2002 in tetrahydrofuran. The fractions obtained were also checked by thin layer chromatography (DC 1: n-hexane-ethyl acetate = 3:1 + 5 % methanol; DC 2: dichloromethane-methanol = 96:4). The lowest molecular weight fraction 6 contained acetylated dodecanol and glucose pentaacetate in approx. 10 %. By ^{1}H-n.m.r. fraction 5 (approx. 60 %) proved to be dodecyl α- and ß-glucopyranoside peracetate (M=517).

Fraction 4 approx. (14 %) obviously contained three main components. By HPLC separation two of these could be isolated, and the structure elucidation revealed dodecyl heptaacetyl disaccharides (M = 805). By ^{1}H n.m.r. the heptaacetate of dodecyl α-isomaltoside and of dodecyl α-nigeroside could be assigned unequivocally. The exclusive formation of α-interglycosidically linked disaccharides is in accordance with the expectation of a Fischer glycosylation. Whereas the occurrence of the 1-6 linkage can be rationalised easily, that of the 1-3 linkage seems to be a surprise and requires mechanistic interpretation. In fractions 3 and 2 (some 16 %) tri-, tetra-, penta-, and higher oligosaccharides are accumulated with molecular weights ranging from 1000 to approx. 1800. These results underline the correct naming of this APG sample as 'hydrophilic'. With an 3.5-fold excess of fatty alcohol mainly mono and disaccharide glycosides result and the degree of polymerisation is comparatively low.

Another sample called 'hydrophilic APG, 2:1' was prepared from the butyl glucoside precursor mixture with dodecanol in the ration 2:1 and otherwise the same conditions as above. Again fraction 6 (approx. 5 %) , fraction 5 (20 %), fraction 4 (approx. 10 %) and fraction 3 (approx. 12 %) are observed and the components assigned as in the (a) sample. However, there are high molecular weight fractions 2 and 1 (M > 2000) and these contain more than 50 % of the total mass. For further analysis the peracetylated material was subjected to a flash-chromatographic separation using dichloromethane - methanol = 96 : 4, and further to a separation on a reversed phase RP-2 column (methanol-water = 85:15). By ^{1}H and ^{13}C n.m.r. as well as DCI mass spectrometry some information on the high molecular components can be obtained. The average molecular weight may be estimated to be around M ~ 4000, and the interglycosidic linkages of the oligosaccharide chain are

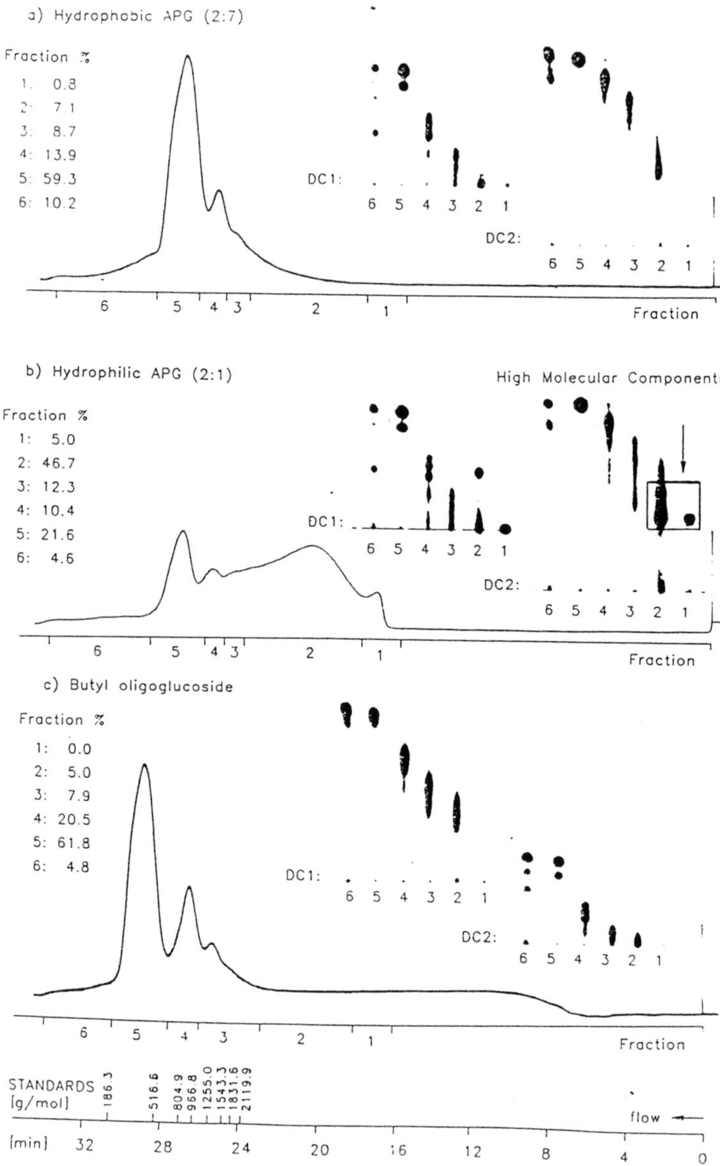

Figure 13 Molecular Distribution and Separation of APG Samples

likely to comprise α and ß, 1 - 6 and α, 1 - 2 linkages. Again this sample prepared with a small excess of dodecanol shows considerable reactions of glucose units leading to a higher degree of polymerisation as evident from these data.

Sample (c) reflects the distribution of compounds observed for the starting material 'butyl oligoglucoside'. The graph immediately resembles that of the hydrophobic APG (a), and also the ratio of fraction 6 to 1, and the t.l.c. observations are comparing very well for (a) and (c). In the preparation an excess of butanol in relation to the sugar is used, and thus this approach is somewhat similar to that in (a), giving this time the butyl glucosides of mono to pentasaccharides with a degree of polymerisation DP ~ 1.8

Functionalised Alkyl Glycosides

Modification of the basic carbohydrate structures allows nature to employ sugar derivatives in a wide range of metabolic processes as well as for special features in plants and animals. In particular the ubiquitous uronic acids and sulfated sugars represent natural anionic surfactants and are found in a number of interesting applications (Fig. 14). Glycosaminoglycans[42] such as the polysaccharides hyaluronate (**37**), chondroitin (**38**), and the anticoagulant heparin (**39** and **40**) form the dominant structures of connective and cartilage tissue in mammals. Plant cell walls contain pectins (**41**)[43] with the characteristic feature to form gels. Among others the galactosyl cerebroside (**42**) may be mentioned which is the glycolipid part of a multilayer membrane structure surrounding nerve fibers.[44]

Uronic Acid Oligosaccharide Glycosides

A simple attractive and well documented[45] approach to uronic acid derivatives is by catalytic oxidation.[46,47] Advantageously employed here was the Adams catalyst under turboagitation at pH 7-8.5 which gave exclusively the terminally oxidised species in yields between 75 and 95 % within 3-7 hours. For example tetradecyl α-maltoside (**36**) gave the 6'-uronic acid glycoside **43**, and correspondingly the dodecyl α-maltotrioside (**13α**) led to the 6"-uronic acid glycoside **44.**

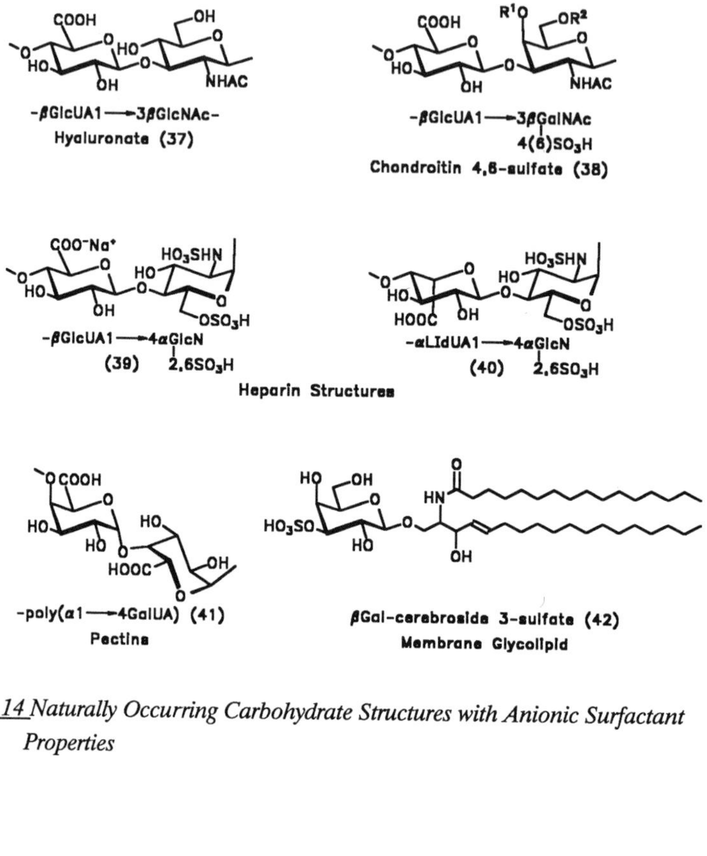

-βGlcUA1→3βGlcNAc-
Hyaluronate (37)

-βGlcUA1→3βGalNAc 4(6)SO_3H
Chondroitin 4,6-sulfate (38)

-βGlcUA1→4αGlcN 2,6SO_3H (39)

-αLIdUA1→4αGlcN 2,6SO_3H (40)

Heparin Structures

-poly(α1→4GalUA) (41)
Pectins

βGal-cerebroside 3-sulfate (42)
Membrane Glycolipid

Figure 14 Naturally Occurring Carbohydrate Structures with Anionic Surfactant Properties

For structural elucidation these derivatives were also peracetylated in the usual manner for studies with ^1H n.m.r. Under these conditions the terminal uronic acid group gives rise to 3,6-lactone formation (**46** and **47**, respectively). Further, a base-catalysed ß-elimination of acetic acid[48] in the terminal sugar led to the unsaturated compounds **45** and **48**, respectively. In contrast to earlier reports on the simultaneous oxidation of both primary hydroxy groups[49] our results are in accord with findings of Dutton[50] and Jayme et al.[51] on the regiospecific oxidation of the terminal hydroxymethylene group in disaccharide glycosides. The regiospecificity may be assumed to be due to extensive intramolecular hydrogen bonding with the 'inner' CH_2OH groups in particular in remaining helix structure elements in maltooligomers.

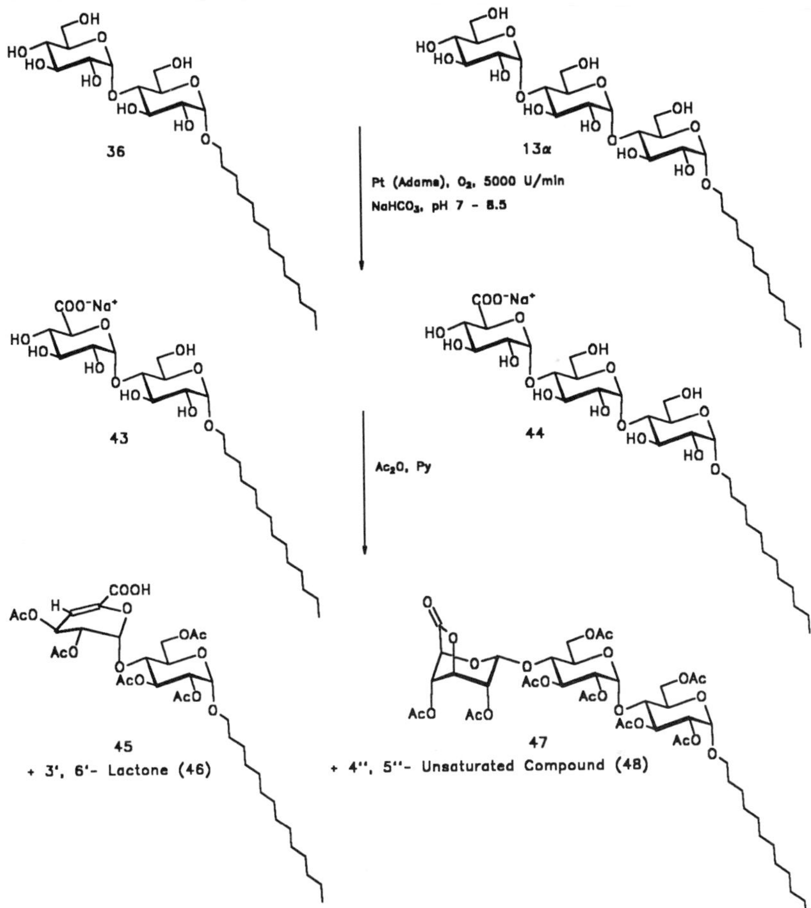

Figure 15 Catalytic Oxidation of Oligosaccharide Glycosides

It was interesting to check the principle feasibility of an enzymatic approach[52,53] for the preparation of alkyl uronates from alkyl glycosides. Uridine-5'-diphosphoglucuronyl transferase (UDPGT, EC 2.4.1.17) in vivo catalyses ß-glucuronylation of a variety of endogenic as well as exogenic substrates prior to elimination from the metabolism. Recently some conjugates of drugs[54,55] could be obtained following this conception. Treatment of the glucuronate donor uridine diphospoglucuronate (UDP-GlcUA, **49**) with dodecanol as the acceptor in the presence of the enzyme UDPGT from bovine liver gave dodecyl ß-glucuronate (**50**) in approx. 70 % yield (Fig. 16). The structure was unequivocally proven by ^{13}C n.m.r. spectroscopy. As with other glucuronides the isolation process may be troublesome, but best results were obtained by chromatography using reversed phase RP-2 columns. Dependent on the future development an alternative formation of special uronates may be envisaged by an enzymatic approach.

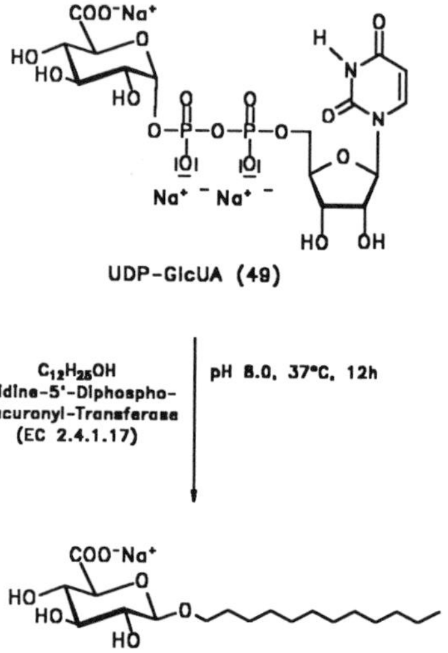

UDP-GlcUA (49)

$C_{12}H_{25}OH$
Uridine-5'-Diphospho-glucuronyl-Transferase
(EC 2.4.1.17)

pH 8.0, 37°C, 12h

50

Figure 16 Enzymatic Preparation of Dodecyl ß-D-Glucuronate (50)

Sulfated Oligosaccharide Glycosides

Another study was initiated to transform the APG model compounds into regioselectively sulfated derivatives directly without extensive protective group chemistry. Based on earlier work[56] sulfation was achieved with the sulfur trioxide-pyridine complex by adjusting reaction parameters such as temperature and molar ratio of starting material and reagent.[57] As depicted in Fig. 17 treatment of dodecyl ß-glucopyranoside (**5**) at -10 °C to room temperature with two molar equivalents of sulfur trioxide-pyridine gave the 6-sulfate **52** in about 50 % yield. By increasing the temperature to 20 °C and doubling the molar amount of SO_3 three regioisomeric sulfates, the 2,6-(**53**), the 3,6-(**54**), and the 4,6-disulfate (**55**) are obtained in 18, 28, and 26 % yield, respectively. By further increase of the molar ration in favour of SO_3 the persulfated glycoside **56** results in almost 70 % yield. The ^{1}H n.m.r.

distribution of negative charge on both sides of the pyran ring system.

Corresponding experiments with dodecyl α-glucopyranoside (**51**) show a rather high yield (78 %) of the 6-sulfate **57** by treatment with a molar excess of SO_3 at -10 °C. In contrast to the reaction of the ß-glucoside treatment of the anomer **51** with four moles of SO_3 at room temperature gives exclusively the 2,6-disulfate **58** in 75 % yield. This amazing regioselectivity may be understood by the observed decreased acidity of the 2-OH group which correlates with an increased nucleophilicity.

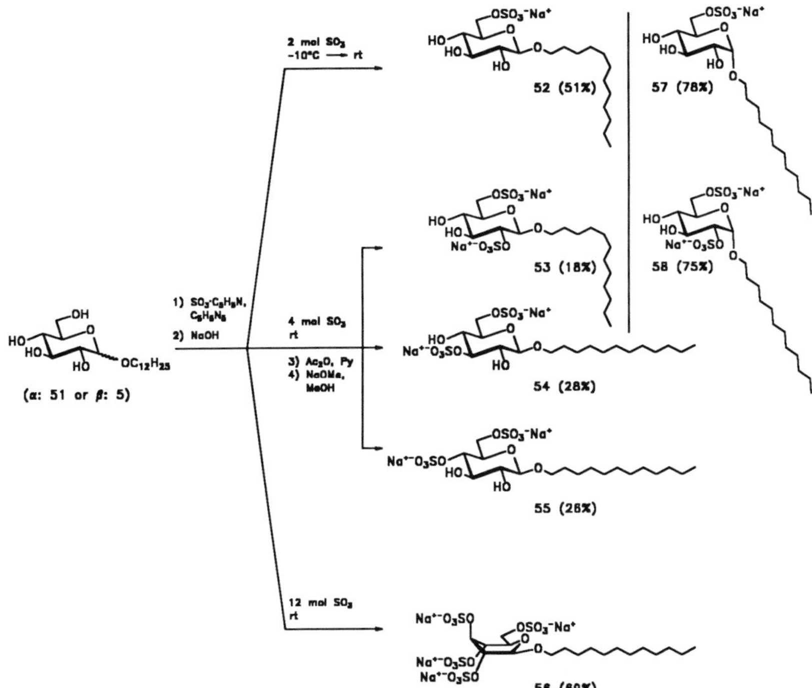

Figure17 Sulfation of Dodecyl α- and β-D-Glucopyranosides

Finally, sulfation experiments of the disaccharide glycoside **30** should give information of an envisaged trend in extending such a reaction to APG samples. By treatment of tetradecyl ß-maltoside with four moles of SO_3 at -10 °C to room temperature both the 6'-(**59**) and the 6-monosulfate (**60**) were obtained in a ratio of about 9:1. Evidently there is observed a predominant esterification at the terminal primary hydroxy group most likely in accordance with steric requirements. However, in contrast to the regiospecificity in the oxidation reaction sulfation gave the 6-ester **60** as well.

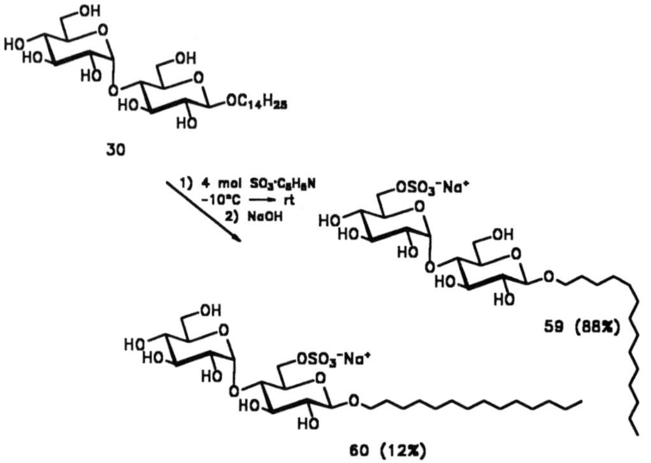

Figure 18 *Sulfation of Tetradecyl β-Maltoside*

Acknowledgement

Support of this work by the **Fonds der Chemischen Industrie** and the **Hüls AG,** Marl, is gratefully acknowledged.

References

1. H.A. Krässig, Schriftenreihe des Fonds der Chemischen Industrie, 1985, 24, 16.

2. J.M. Tedder, A. Nechvatal, A.W. Murray and J. Carnduff, 'Basic Organic Chemistry', Part 4, John Wiley & Sons, London, 1972.

3. Deliberately in this figure the important development of polymeric materials by modification of natural carbohydrate polymers or by de novo synthesis from carbohydrate building blocks was not incorporated.

synthesis from carbohydrate building blocks was not incorporated. For leading references compare e.g. F. Bachmann and J. Thiem, Makromol. Chem., 1991, 192, 2163.

4. 'Towards a carbohydrate-based chemistry', Symposium Proceedings, Amiens, 23-26 Oct. 1989 (Eds.: D.C. Ellwood et al.), EUR 12757, Luxembourg 1990.

5. H. Zoebelein, Chem. Ind. (Düsseldorf), 1988, 3, 30.

6. I.D. Robb in 'Industrial Application of Surfactants II' (Ed.: D.R. Karsa), Royal Society of Chemistry, Cambridge, 1990, p. 22.

7. F. Wagner and S. Lang, Proceedings 2nd World Surfactants Congress, Paris 1988, Vol. II, p. 71.

8. C. Syldatk, U. Matulovic and F. Wagner, Biotech-Forum , 1984, 1 (3/4), 58.

9. N. Weber and H. Benning, Chem. Phys. Lipids, 1982, 31, 325.

10. N. Weber and H. Benning, J. Nutr., 1984. 114, 247.

11. J.R. Hurford in 'Development in Food Carbohydrate-2' (Ed.: C.K. Lee), Applied Science Publ. Ltd., London, 1980, Chap. 7, p. 327.

12. E. Fischer, Ber. Dtsch. Chem. Ges., 1885, 28, 1145.

13. A.F. Bochkov and G.E. Zaikov, 'Chemistry of the O-Glycosidic Bond', Pergamon Press, Oxford, 1979.

14. W. Koenigs and E. Knorr, Ber. Dtsch. Chem. Ges., 1901, 34, 957.

15. P. Rosevear, T. Vanaken, J. Baxter and S. Ferguson-Miller, Biochemistry, 1980, 19, 4108.

16. D.E. Koeltzow and A.D. Urfer, J. Am. Oil Chem. Soc., 1984, 61, 1651.

17. D.E. Koeltzow, J. Carbohydr. Chem, 1985, 4, 125.

18. Th. Böcker, Diplomarbeit, Univ. Münster, 1987.

19. J. Thiem, Th. Böcker, P. Hofmann, H. Lüders and G. Schreier, Proceedings 2nd World Surfactants Congress, Paris 1988, Vol. I, p. 197.

20. H. Paulsen, Angew. Chem., 1982, 94, 184; Angew. Chem. Int. Ed. Engl., 1982, 21, 155.

21. P.W. Austin, F.E. Hardy, J.G. Buchanan and J. Baddiley, J. Chem. Soc., 1963, 5350.

22. A. Neuberger and B.M. Wilson, Carbohydr. Res., 1971, 17, 89.

23. G. Zemplen, Ber. Dtsch. Chem. Ges., 1929, 62, 985.

24. B. Lindberg, Arkiv Kemi, Mineral. Geol. B., 1944, 18, 1.

25. R. U. Lemieux and W.P. Shyluk, Can. J. Chem., 1953, 31, 528.

26. J. Banoub and D.R. Bundle, Can. J. Chem., 1978, 57, 2085.

27. T. Vanaken, S. Foxall-Vanaken, S. Castleman and S. Ferguson-Miller, Methods Enzymol., 1986, 125, 27.

28. K. Bock and H. Thogersen, Ann. Rev. NMR Spectrosc., 1982, 13, 1.

29. K. Bock and C. Pedersen, Adv. Carbohydr. Chem. Biochem., 1983, 41, 27.

30. K. Bock, C. Pedersen and H. Pedersen, Adv. Carbohydr. Chem. Biochem., 1984, 42, 193.

31. M. Forsgren, P.-E. Jansson and L. Kenne, J. Chem. Soc., Perkin Trans.1, 1985, 2383.

32. I. Backmann, B. Erbing, P.-E. Jansson, and L. Kenne, J. Chem. Soc., Perkin Trans. 1,1988, 889.

33. M. Hesse, H. Meier and B. Zeeh, 'Spektroskopische Methoden in der Organischen Chemie', 2. Ed., Thieme, Stuttgart 1984.

34. R.J. Ferrier and N. Prasad, J. Chem. Soc. (C), 1969, 571.

35. R.J. Ferrier, Adv. Carbohydr. Chem, 1965, 20, 67.

36. R.J. Ferrier, Adv. Carbohydr. Chem. Biochem., 1969, 24, 199.

37. B. Iselin and T. Reichstein, Helv. Chim. Acta, 1944, 27, 1146.

38. W.N. Haworth, E.L. Hirst and R.J.W. Reynolds, J. Chem. Soc., 1934, 303.

39. Th. Böcker and J. Thiem, Tenside Surf. Det., 1989, 26, 318.

40. P. Lecomte du Noüy, J. Gen. Physiol., 1919, 1, 521.

41. K. Shinoda, T. Yamaguchi and R. Hori, Bull. Chem. Soc. Jpn., 1961, 34, 237.

42. J.F. Kennedy (Ed.), 'Carbohydrate Chemistry', Oxford Science Publications, Clarendon Press, Oxford 1988, Chap. 8.

43. G.O. Aspinall in 'The Carbohydrates' (Eds.: W. Pigman et al.), 2n Edn., Vol. IIB, Academic Press, New York 1970, p. 515.

44. D.E. Vance in 'Biochemistry' (Ed.: G. Zubay), Addison-Wesley, Reading, Massachusetts 1983, p. 528.

45. C.L. Mehltretter, B.H. Alexander, R.L. Mellies and C.E. Rist, J. Am. Chem. Soc., 1951, 73, 2424.

46. K. Heyns and H. Paulsen, Adv. Carbohydr. Chem., 1962, 17, 169.

47. K. Heyns, H. Paulsen, G. Rüdiger, and J. Weyer, Fortschr. Chem. Forsch., 1968, 11, 285.

48. J. Kiss, Adv. Carbohydr. Chem. Biochem., 1970, 29, 230.

49. D. Abbott and J. Weigel, J. Chem. Soc., 1965, 5157.

50. G.G.S. Dutton and K.N. Slessor, Can. J. Chem., 1964, 42, 1110.

51. G. Jayme and W. Demmig, Chem. Ber., 1960, 63, 356.

52. G.M. Whitesides and C.-H. Wong, Angew. Chem., 1985, 97, 617; Angew. Chem. Int. Ed. Engl., 1985, 24, 617.

53. R. Stiller and J. Thiem in 'Jahrbuch Biotechnologie', Vol 3, (Eds.: P.Präve et al.), C. Hanser, München, 1990, p. 101.

54. R. Neidlein, M. Wu and H.G. Hege, Arzneim.-Forsch./Drug Res., 1988, 38, 1257.

55. D. Gygax, M. Hammel, R. Schneider, E.G. Berger and H. Stierlein in 'Enzymes in Carbohydrate Synthesis' (Eds.: M.D. Bednarski and E.S. Simon), ACS Symp. Ser. 466,ACS, Washington, DC 1991, Chap. 6, p. 79.

56. N. Miam, C.E. Anderson and P.W. Kent, Biochem. J., 1979, 181, 387.

57. V. Vill, Th. Böcker, Th. Lindhorst and J. Thiem, Carbohydr. Res., in press.

Applications of 3-Alkyloxypropylamines and their Derivatives

M.L. Durrant

EXXON CHEMICAL LTD., EXXON CHEMICAL RESEARCH CENTRE, P.O. BOX 1, ABINGDON, OXON OX13 6BB, UK

1. INTRODUCTION

Primary aliphatic amines of C_{10}-C_{20} carbon chain length are an important class of industrial chemicals. They are used widely as chemical building blocks for surfactants, biocides, polymer additives, fuel and lubricant additives, and in applications as diverse as mineral processing, corrosion inhibition, road construction and water treatment. Such amines can be used in this diversity of applications because the extensive chemistry of the amino function allows preparation of useful derivatives, each endowed with very special properties. Amongst these, the cationic surfactant properties of amines and amine derivatives are paramount and a recurring theme in their applications is substantivity. The strong adsorption of cationic surfactants on negatively charged surfaces manifests itself in effects such as improved rinsing, softening, corrosion inhibition and biocidal activity.

Amines in this molecular weight range are available from both vegetable and animal fat sources and petroleum feedstocks. The ammonolysis of fatty acids is used to manufacture linear C_8-C_{22} amines although in practice, the most important commercial grades range between C_{12} and C_{18}. The second major source of feedstock for surfactant-range amines are higher olefins derived from petroleum. Potentially, amines can be derived from these olefins in a number of ways: by treatment with hydrogen halides and then ammonia or from alcohols made by hydroformylation of the olefin - the OXO process. These alcohols can be used to make amines by either direct treatment with ammonia or treatment with halide and subsequent ammonolysis. Commercially, one of these latter two routes is used to make amines consisting of a mixture of branched and linear C_{13} (70%) and C_{15} (30%) alkyl groups.

3-Alkyloxypropylamines (1) are an unusual and versatile class of synthetic amine made by the reaction of an alcohol with acrylonitrile with subsequent hydrogenation. This class of amines is unusual in that many types of alcohol, both fat- and petroleum-based, can be used as raw material. This permits the preparation of amines, and hence amine derivatives, which cover a wide range of carbon chain length and differing surfactant properties.

$$R\text{-}O\text{-}CH_2\text{-}CH_2\text{-}CH_2\text{-}NH_2$$

(1)

2. SYNTHESIS

Aliphatic alcohols (2) will undergo 1,4 conjugate addition with acrylonitrile (3) in the presence of base and at a temperature of 80-90°C. The reaction is also termed cyanoethylation[1].

The intermediate ether nitrile (4) is then hydrogenated over Raney nickel catalyst at a gas pressure of 300-400 psi and a temperature of 90-150°C. The yield is essentially quantitative from the alcohol with the only by-product being negligible amounts (< 1%) of acrylonitrile polymer.

$$\underset{(2)}{R\text{-}OH} + \underset{(3)}{CH_2{=}CHCN} \text{ -------> } \underset{(4)}{R\text{-}O\text{-}CH_2\text{-}CH_2\text{-}CN}$$

$$\underset{(4)}{R\text{-}O\text{-}CH_2\text{-}CH_2\text{-}CN} + 2H_2 \text{ --------> } \underset{(1)}{R\text{-}O\text{-}CH_2\text{-}CH_2\text{-}CH_2\text{-}NH_2}$$

Figure 1 Synthesis of 3-Alkyloxypropylamines

The whole reaction process from alcohol (2) to 3-alkyloxypropylamine (1) is conducted in one vessel without isolation of the intermediate ether nitrile (4).

3. PHYSICAL PROPERTIES AND FEATURES

Variety of Hydrophobe

The reaction scheme shown in Figure 1 can make use of almost any aliphatic alcohol as raw material. In this fashion, the amino function and the opportunities it offers for derivative synthesis can be allied to hydrophobes of widely differing properties. Alkyloxypropylamines can thus offer interesting opportunities for rational surfactant design. Table 1 shows the variety of R group in commercially available 3-alkyloxypropylamines.

Table 1 Commercially Available 3-Alkyloxypropylamines

Alcohol Type	R-Group Carbon Chain Length
Zeigler	C_{12} (n-)
Alpha-Olefin OXO	C_{9-11}, C_{13-15} (n-/i-)
Propylene OXO	C_{6-13} (i-)
Natural	C_8-C_{18} (n-)
Guerbet	C_{16},C_{20},C_{26} (i-)
2-Ethylhexanol	C_8 (i-)
3,5,5 Trimethylhexanol	C_9 (i-)

Liquidity

A key property of 3-alkyloxypropylamines is their much greater liquidity when compared to fatty amines of a similar molecular weight. Whilst fatty amines are normally waxy or pasty solids at ambient temperature, alkyloxypropylamines are flowing, easily pourable liquids (see Table 2). They retain this liquidity to temperatures well below freezing. This feature can be of great value in handling, processing and synthesis, often eliminating the need for heating or dissolution in solvent as can be required for waxy amines.

Table 2 Pour Points of Alkyloxypropylamines and Fatty Amines

$$R\text{-}O\text{-}CH_2\text{-}CH_2\text{-}CH_2\text{-}NH_2$$

(1)

Amine	Typical Molecular Weight	Pour Point/°C
Laurylamine	185	24-29
(1), R = Isohexyl	165	-33
Cocoamine	200	13-17
(1), R = Isodecyl	229	< -38
Tallowamine	262	31-38
(1), R = Isotridecyl	274	< -38

4. APPLICATIONS

Applications for alkyloxypropylamines and their derivatives are legion and the review of the patent literature given hereunder describes the principal applications as judged by the quality and quantity of patent activity.

3-Alkyloxypropylamines

Ore Flotation Reagents. Ore flotation is a process for separating finely ground valuable minerals from their associated gangue and depends on the relative wettability of the surfaces of the various mineral species in the crushed ore after suitable modification. In the case of negatively charged mineral surfaces such as those found on silicates, the addition of cationic surfactants renders the surface hydrophobic and more readily wetted by air bubbles. 3-Alkyloxypropylamines and their partial acetate derivatives have been extensively used in the beneficiation of several minerals including kaolin[2] and chalcopyrite copper sulphide[3]. The largest commercial application for alkyloxypropylamines is as the collector for silicate gangue from iron oxide[4-6] and certain phosphate ores, where better yields and high selectivity combine with extreme ease of handling (the acetate derivatives are all liquid) to make alkyloxypropylamines more cost effective than their fat-based counterparts. Several thousand tonnes of alkyloxypropylamines are being used worldwide in this application.

Epoxy Curing Agents. Several 3-alkyloxypropylamines have been used as hardening agents or as part of a hardening system for epoxide resins. When 3-(n-octyloxy)- and 3-(n-dodecyloxy)propylamines were used as curing agents for epoxy mouldings, the resultant mouldings showed good flexibility[7]. Epoxy coatings hardened with 3-(2-ethylhexyloxy)propylamine have been used in asphalt road dressings, providing a top coat with flow properties and flexibility comparable to the asphalt. Apart from the advantage this gave in the handling of asphalt, such coatings were reported to give good skid-proofing[8]. N,N-Dimethyl derivatives of 3-(n-dodecyloxy)- and 3-(1-methyldecyloxy)propylamines improved the pot-life of epoxy resin adhesives compared to hardeners based on more volatile amines such as N,N-dimethyldiaminopropane[9].

Fuel and Lubricant Additives. The liquidity, corrosion inhibition and biocidal properties of alkyloxypropylamines have been used to good effect in fuel and lubricant additive applications. 3-(n-Octyloxy)-, 3-(n-dodecyloxy)- and 3-(2-ethylhexyloxy)- propylamines are useful oil-soluble neutralisers in petroleum refining processes. When crude oil was distilled in the presence of these amines, no precipitation of neutralised salts were observed in the distillation column. In this way, the build-up of acidic components could be controlled but without the problem of precipitation, as was seen when 3-(methoxy)propylamine was used instead[10].

The addition of 0.02 wt% of 3-(isotridecyloxy)propylamine to ethanol inhibits the corrosive properties of alcohols toward ferrous metals. This is particularly useful in gasoline blends where ethanol corrosion can occur[11].

3-(2-Ethylhexyloxy)propylamine shows excellent rust and fungal growth inhibition when added to emulsion-based cutting fluids[12]. A variety of 3-alkyloxy- propylamines have been shown to be good inhibitors of slime formation in water-based lubricants[13].

Miscellaneous. Alkyloxypropylamines have been used in a variety of other applications. Notably, they have been used as fungicides for rice crops[14], cucumber and tomato plants[15] and timber[16]. 3-(n-Dodecyloxy)propylamine has been used as the principal anti-bacterial agent in an eye lotion for the treatment of conjunctivitis[17]. 3-(n-Dodecyloxy)propylamine is an effective dispersant for coal in fuel oil, stabilising the dispersion and leading to minimal separation on storage[18]. 3-(Isohexyloxy)- and 3-(isodecyloxy) propylamines have been used as foaming aids in the manufacture of mineral wool panels used for sound and thermal insulation[19].

Alkyloxypropylamines with linear and branched C_8-C_{10} alkoxy groups improved the adhesion of bitumen and wet paving aggregates[20]. The addition of mono- or diethanolamine attenuated the adhesion effect.

3-Alkyloxypropylamine Derivatives

Surfactants. Whilst alkyloxypropylamines themselves have useful and interesting detergent properties, this class of amine has been used extensively in the preparation of other surfactant molecules. Most recently, it has been reported that lower ethoxylates of alkyloxypropylamines derived from ethoxylated alcohols boost detergency in the laundering of soiled cotton and polyester-cotton fabrics containing oily soils such as lipstick and make-up[21]. 3-(i-Tridecyloxy)- and 3-(n-octadecyloxy)propylamines have been reacted with maleic anhydride-acrylic acid copolymers (MW 45,000), producing a mixture of neutralised and amidated carboxylate groups. These polymers were reported to be detergency improvers and greying inhibitors for laundry detergent compositions[22].

3-(n-Dodecyloxy)propylamine has been used in the preparation of N,N-disubstituted beta-aminopropionic acids, surfactants that are particularly effective in the process of rendering leather hydrophobic[23].

N-(Alkyloxypropyl) sulphosuccinamates are very soluble in aqueous solution and 35% solutions are clear and stable at room temperature. By contrast, sulphosuccinamates derived from tallow amine need to be warmed to 30-35°C to maintain a clear, non-viscous solution at this concentration. N-(Alkyloxypropyl) sulphosuccinamates derived from 3-(n-dodecyloxy)-, 3-(n-tetradecyloxy)- and 3-(n-octadecyloxy)propylamine have been used as surfactants with good foaming properties in carpet shampoos[24].

Aminosuccinates based on alkyloxypropylamines have been used as emulsifiers in the emulsion polymerisation of butadiene. The resultant latexes have improved shear stability and coagulation behaviour[25].

Aluminosilicate clay thickeners are used in many applications where precise viscosity control is required. Whilst these clays are easily dispersed in water, a cationic surfactant must be added in order to render the clay organophilic if viscosity control is required in organic systems such as bituminous coatings or alkyd resin paints. The fully neutralised acetate salt of 3-(i-decyloxy)propylamine will reduce the interfacial tension between attapulgite clay and asphalt, for example, so allowing the clay to remain dispersed and acheive optimum thickening in the gel.

When compared to the mono- and dialkyl quaternary ammonium salts traditionally used as attapulgite clay modifiers, much less 3-(i-decyloxy)propylamine acetate is needed to acheive stable, optimum gel viscosities[26]. The liquidity and non-flammability of the alkyloxypropylamine salt are also important advantages.

Anti-Static Additives. Amine ethoxylates are used widely in plastics to prevent static electricity build-up and dust collection on packaging films, flooring products, furniture compounds, etc. They are compounded with the polymer resin and then migrate to the polymer surface to provide a conducting layer. The liquidity and lower pour points of alkyloxypropylamine ethoxylates provides easier dispersion and blending characteristics in polymer resins. Ethoxylate derivatives of a variety of alkyloxypropylamines with both branched and linear alkyloxy groups exhibit excellent anti-static properties for polystyrene[27], polyethylene, polypropylene and polybutylene mouldings[28].

Control of static charge build-up can also be effected by application of an external coating to a processed polymer and solutions of quaternary ammonium salts are commonly used. This control of static charge is of particular importance during the processing of synthetic fibres. The 2 mole ethoxylates of 3-(n-dodecyloxy)- and 3-(n-tetradecyloxy)propylamines were each mixed with triethanolamine (9:1) and each mixture quaternised with diethylsulphate. Each resultant quaternary ammonium salt mixture was liquid at room temperature and 0.5% aqueous solutions of these products showed excellent static charge dissipation when applied to polyester yarns[29].

Fuel and Lubricant Additives. Several alkyloxypropylamines have been used in the synthesis of nitrogen-containing fuel and lubricant additives. Such additives are used to control problems such as corrosion, injector valve and carburettor fouling. It is the key properties of substantivity and surfactancy that make amine and amine derivatives effective in their control.

3-(i-Tridecyloxy)- and 3-(n-octadecyloxy)propylamine were used in the synthesis of

N,N'-bis(3-alkyloxypropyl)asparagines, derivatives which were shown to improve the carburettor detergency and anti-corrosive properties of hydrocarbon fuels[30]. Polyisobutyl tartarimides in which the imide nitrogen was derived from 3-(2-ethylhexyloxy)-, 3-(i-undecyloxy)- and 3-(i-tetradecyloxy)propylamines showed good corrosion inhibition and dispersant properties for fuels and lubricants[31]. 3-(2-Ethylhexyloxy)propylamine has been incorporated in several tetramides based on EDTA[32]. Such tetramides were shown to be good detergents for inlet valves and carburettors.

Dihydrouracils were manufactured as deposit modifiers for lubricant oils by the reaction of polyisobutenyl-substituted N-hydroxysuccinimides with 3-(i-decyloxy)propylamine. When added at 2% by weight in a fully formulated crankcase oil, the adhesion of oxidised engine deposits was reduced[33].

Dyes and Pigments. Alkyloxypropylamines have been used to modify the dispersion characteristics of dyestuffs and improve solubility of the dyestuff in oil- or resin-based systems.

3-(2-Ethylhexyloxy)propylamine has been used in the synthesis of anthroquinone-based dyes which have particular application as providing colour in plastics and liquid crystals[34]. Various linear and branched alkyloxypropylamines have been used in the synthesis of both phthalocyanine[35] and azo[36] dyes used in the preparation of aqueous ink compositions for jet-printing. These ink formulations were reported to have good stability on storage, prevent clogging of the jetting nozzle, have excellent image clarity and good water resistance.

3-(Ethyloxy)- and 3-(2-ethylhexyloxy)propylamines were used to prepare perylene-3,4,9,10-tetracarboxylic acid diimide dyes which were reported as giving fast orange shades to PVC-styrene block co-polymers[37].

Miscellaneous. 3-(n-Dodecyloxy)propylamine has been used in the synthesis of substituted propionic acid amides which have useful bactericidal and fungicidal properties[38]. The same workers also reported significant anti-microbial activity from butyric acid amides[39] and N-substituted alkylamines[40] using this alkyloxypropylamine.

The sulphite derivative of 3-(n-octyloxy)propylamine showed useful corrosion inhibition behaviour toward ferrous surfaces in brine-crude oil mixtures[41]. This property is beneficial in different areas of the oil recovery and petroleum industry.

5. SUMMARY

3-Alkyloxypropylamines, prepared by the cyanoethylation of aliphatic alcohols, are typified by high liquidity when compared to amines of comparable molecular weight derived from agricultural sources. In addition, the cyanoethylation reaction permits a wide range of hydrophobe to be attached to the propylamino moiety by selective choice of alcohol raw material. This can offer advantages in the design of surfactant derivatives.

3-Alkyloxypropylamines show interesting surfactant properties in their own right and they and their derivatives have been used in all the major industrial applications normally associated with fatty amines. However, in addition to classical cationic surfactant properties, 3-alkyloxypropylamines and their simple derivatives offer high solubility in oil or other organic systems, ease of handling and low pour points.

REFERENCES

1. H.A.Bruson, Org.React. 1949, 5, 79-135.
2. H.Nakazawa, Sumitomo Metal Mining Co.Ltd., JP 63/123454 (1988).
3. (i) R.Klimpel, Dow Chemical Co., Eur.Pat.Appl.EP 174866 (1986).
   (ii) R.Klimpel, Dow Chemical Co., US 4797202 (1989).
4. D.Shaw, Exxon Research and Engineering Co., Fr.Demande FR 2489714 (1982).
5. D.Shaw, Exxon Research and Engineering Co., US 4319987 (1982).
6. J.-L. Polgaire, Ceca SA, Braz.Pedido PI BR 77/6941 (1978).
7. T.Tomikuda, Ajinmoto Co.Inc., JP 51/28900 (1976).
8. T.Doi, Kao Soap Co.Ltd.,Ger.Offen. DE 2018954 (1970).
9. H.Zondler, Ciba-Geigy AG, Eur.Pat.Appl.EP 17623 (1980).
10. K.Matsubara, Kurita Water Industries Ltd., JP 61/34095 (1986).
11. R.Sung, Texaco Inc., US 4321060 (1982).

12. Yushiro Chemical Industry Co.Ltd., JP 60/38495 (1985).
13. H.Kitamura, Kawasaki Steel Corp., JP 62/41293 (1987).
14. E.Yoshinaga, Kumiai Chemical Industry Co.Ltd., JP 48/7327 (1973).
15. G.Lamb, American Cyanamid Co., US 3291683 (1966).
16. Katayama Chemical Works Co.Ltd., JP 55/143902 (1980).
17. G.Andermann, Laboratoires P.O.S. SA, Eur.Pat.Appl. EP 67084 (1982).
18. K.Sakuma, Lion Fat and Oil Co.Ltd., JP 53/82810 (1978).
19. D.Izard, USG Interiors Inc., PCT Int.Appl.WO 88/5100 (1988).
20. (i) M.Hellsten, Mo och Domsjo AB, US 4038102 (1977).
(ii) M.Hellsten, Mo och Domsjo AB, Swed.SE 354478 (1973).
21. R.Bunte, Henkel K-G.a.A., Ger.Offen. DE 3832589 (1990).
22. W.Denzinger, BASF A.-G., Ger.Offen. DE 3528460 (1987)
23. K.Dahmen, Chemische Fabrik Stockhausen GmbH, Ger.Offen DE 3717961 (1988).
24. J.Faulks, Albright & Wilson Ltd., GB 1306066 (1973).
25. U.Hendricks, Bayer A.-G., Ger.Offen. DE 3509822 (1986).
26. (i) Engelhard Corporation Laboratory Study, "Optimum Surfactants for Attagel 36 Asphalt Thickener", 1987.
(ii) Engelhard Corporation Laboratory Study, "Attagel 50 Thickener: An Effective Organoclay Substitute in Solvent-Based Systems", 1991.
27. K.Rombusch, Chemische Werke Huels A.-G., Ger.Offen. DE 1258083 (1968).
28. K.Rombusch, Chemische Werke Huels A.-G., Ger.Offen. DE 1242861 (1967).
29. R.Garst, National Distillers and Chemical Corp., Ger.Offen. DE 3501293 (1985).
30. W.Cummings, Texaco Inc., US 4144036 (1979).
31. D.Barrer, Lubrizol Corp., US 4237022 (1980).
32. K.Oppenlaender, BASF A.-G., Ger.Offen. DE 3500709 (1986).
33. K.Chou, Texaco Inc., US 4329239 (1982).
34. (i) D.Matsunaga, Nippon Kayaku Co.Ltd., JP 62/54761 (1987).
(ii) D.Matsunaga, Nippon Kayaku Co.Ltd., JP 61/221264 (1986).
35. H.Kawashita, Taoka Chemical Co.Ltd., Eur.Pat.Appl. EP 196901 (1986).
36. H.Kawashita, Taoka Chemical Co.Ltd., Eur.Pat.Appl. EP 194885 (1986).
37. F.Graser, BASF A.-G., Ger.Offen. DE 2451782 (1976).
38. W.Gerhardt, Henkel K.-G.a.A., Ger Offen. DE 3343804 (1985).

39. W.Gerhardt, Henkel K.-G.a.A., Ger Offen. DE 3320137 (1984).
40. W.Gerhardt, Henkel K.-G.a.A., PCT Int.Appl. WO 84/4919 (1984).
41. A.Larsen, Kema Nobel AB, UK Pat.Appl. GB 2011383 (1979).

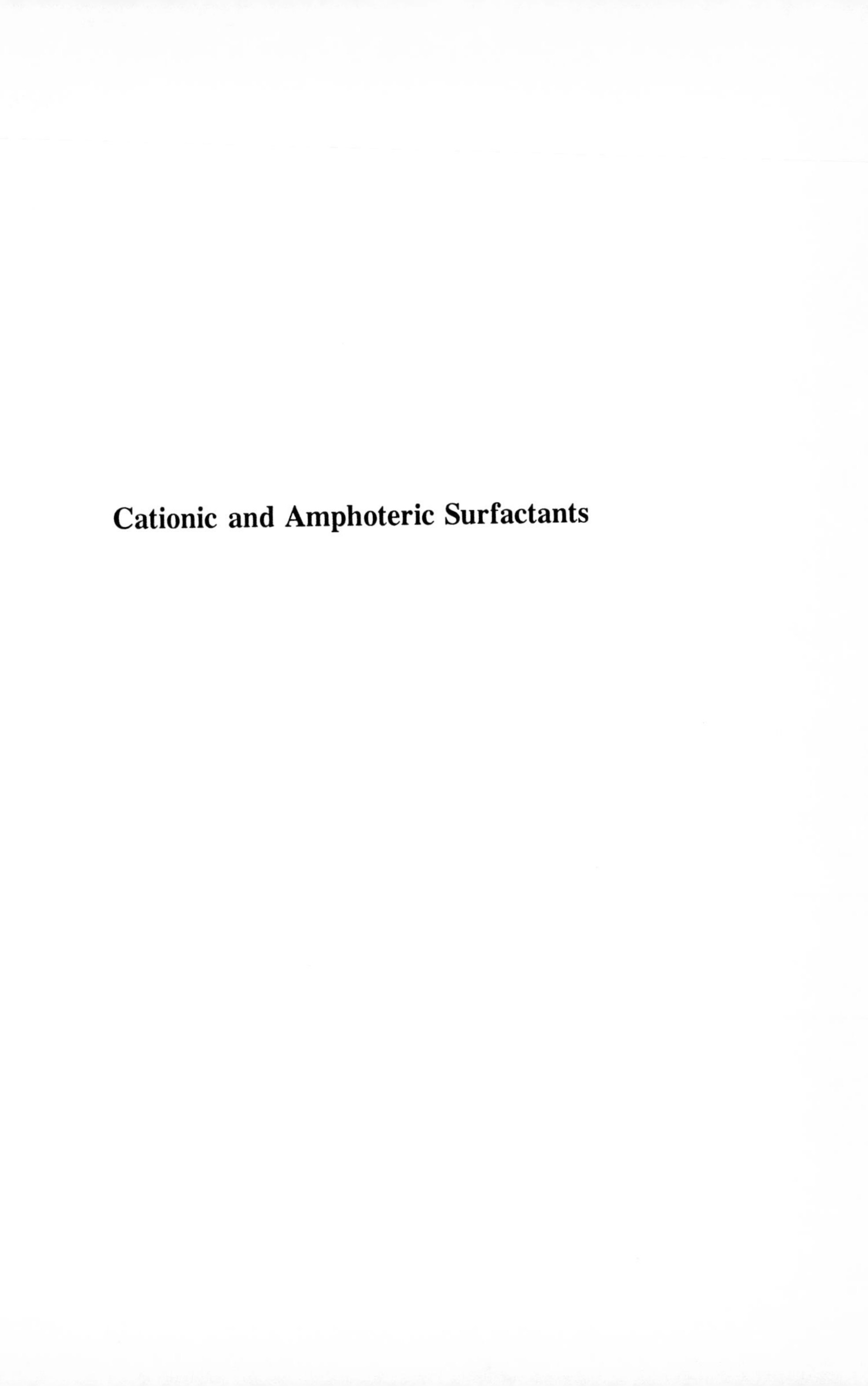

Cationic and Amphoteric Surfactants

Industrial Applications of 2-Alkyl Imidazolines and their Derivatives

F.B. Richardson

THOMAS SWAN & CO. LTD., CROOKHALL, CONSETT, CO. DURHAM DH8 7ND, UK

1 INTRODUCTION

This paper discusses the history, methods of preparation, structure, reactions and applications of 2-alkyl-imidazolines, which hold a rather unique place in the field of cationic surfactants. Although these materials have been around for a long time they still enjoy new developments in both their chemistry and applications. This trend is likely to continue in view of the low toxicity associated with many of the 2-alkyl imidazolines and their derivatives.

II HISTORY

Although the history of 2-alkyl imidazolines dates back to at least 1888, it is during the last fifty years that the potential applications of these compounds have been fully appreciated and exploited. 2-methyl imidazoline was first prepared by Hofmann in 1888 by heating N,N'-diacetylene diamine in a stream of dry hydrogen chloride[1]. The reaction is shown in Figure 1. The important developments in the chemistry and applications commenced in the mid 1930's with claims of excellent wetting, dispersing, cleaning and foaming properties. Figure 1 serves to illustrate many of the developments made during this period including the preparation of quaternary, alkoxy, alkylated, phosphonated, sulphonated and carboxy derivatives[2].

Most 2-alkyl imidazolines will show a level of performance in the applications referred to in this paper. The ultimate performance in a specific application is achieved by careful selection of the main substituents.

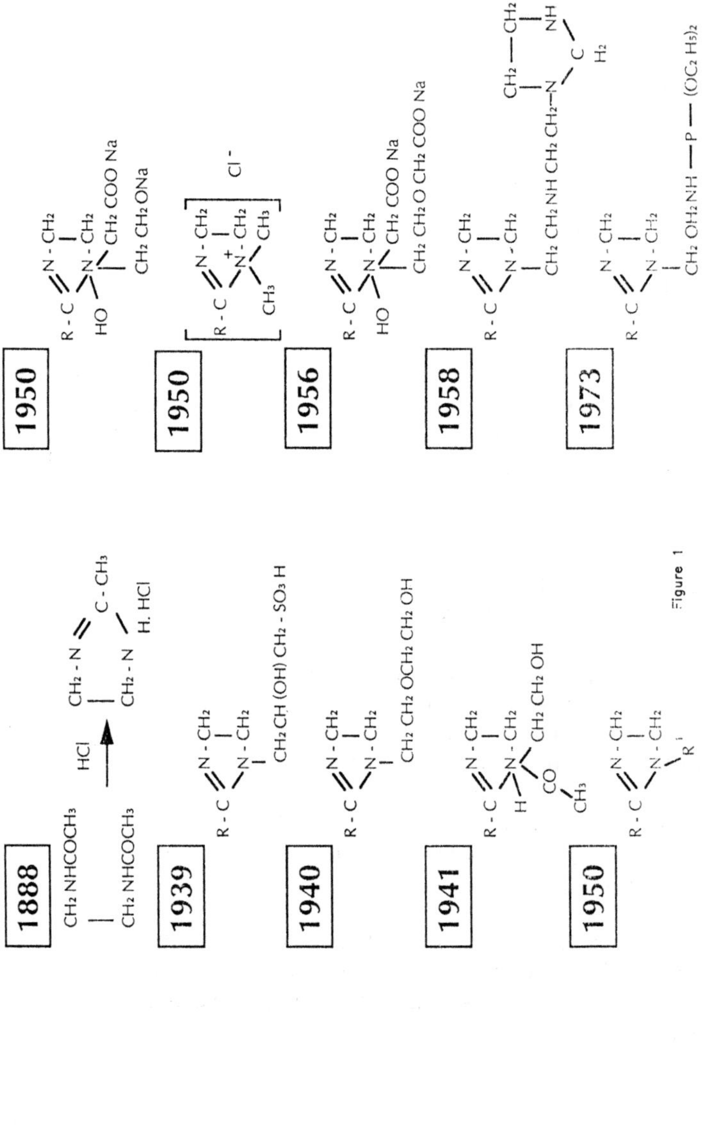

1888

$CH_2NHCOCH_3$ | $CH_2NHCOCH_3$ —HCl→ $CH_2-N=C-CH_3$ / CH_2-N H. HCl

1939

R - C (N - CH_2, N - CH_2) — $CH_2CH(OH)CH_2-SO_3H$

1940

R - C (N - CH_2, N - CH_2) — $CH_2CH_2OCH_2CH_2OH$

1941

R - C (N - CH_2, N - CH_2), H, CH_2CH_2OH, CO, CH_3

1950

R - C (N - CH_2, N - CH_2) — R^1

1950

R - C (N - CH_2, N - CH_2), HO, CH_2COONa, CH_2CH_2ONa

1950

$[R-C(N-CH_2, N^+-CH_2), CH_3, CH_3]$ Cl^-

1956

R - C (N - CH_2, N - CH_2), HO, CH_2COONa, $CH_2CH_2OCH_2COONa$

1958

R - C (N - CH_2, N - CH_2) — $CH_2CH_2NHCH_2CH_2-N$ (CH_2-CH_2, NH, C H_2)

1973

R - C (N - CH_2, N - CH_2) — $CH_2OH_2NH-P-(OC_2H_5)_2$

Figure 1

Figure 1 HISTORY OF 2-ALKYL IMIDAZOLINES

III PREPARATION METHODS

For those interested in an in depth survey of the various methods available for the preparation of 2-alkyl imidazolines, then the article by Ferm and Reibsomer is recommended for further reading[3].

Generally imidazoline syntheses use ethylene diamine or one of its acyl or alkyl derivatives as the starting material. The most widely used methods involve the ring closure of 1:2 diamines or 1:2 diamine derivatives, with carboxylic acids or derivatives of these acids or difunctional carboxylic acids.

Two of the more important preparations are based on the reaction between a fatty acid containing from 12-18 carbons and diethylene triamine or hydroxy ethyl ethylene diamine. These preparations are given in Eqs 1 and 2 shown in Figure 2.

In order to produce high yields of high purity products, special attention must be paid to the molar ratio of reactants, reaction temperature of each stage and applied vacuum during the cyclisation stage. Failure to do this will inevitably lead to undesirable products being present in the final mixture. In the case of Eq 1 these products will be:-

R—C(=N—CH_2)(—N—CH_2) with CH_2CH_2NHCOR on N — amido imidazoline

$RCONHCH_2CH_2NHCH_2CH_2NHCOR$ — diamide — <u>Eq 1</u>

In the case of Eq 2 these products will be:

R—C(=N—CH_2)(—N—CH_2) with CH_2CH_2OOCR on N — ester imidazoline

$RCONHCH_2CH_2NHCH_2CH_2OOCR$ — amido ester — <u>Eq 2</u>

Reactions of diethylene triamine with fatty acids

$RCOOH + NH_2CH_2CH_2NHCH_2CH_2NH_2$

↓

$RCO\bar{O} + \overset{+}{N}H_3CH_2CH_2NHCH_2CH_2NH_2$

<150°C ↓ xs amine

$RCONHCH_2CH_2NHCH_2CH_2NH_2 + H_2O$

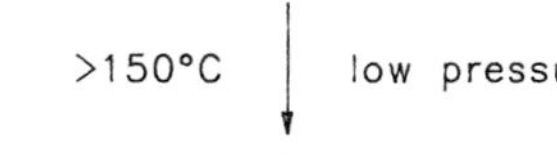

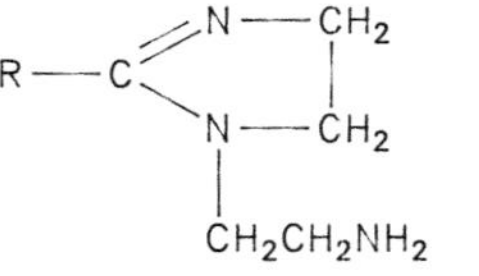

Equation 1

Reactions of hydroxy ethyl ethylene diamine with fatty acids

$RCOOH + NH_2CH_2CH_2NHCH_2CH_2OH_2$

↓

$RCO\bar{O} + \overset{+}{N}H_3CH_2CH_2NHCH_2CH_2OH$

<150°C ↓ xs amine

$RCONHCH_2CH_2NHCH_2CH_2OH + H_2O$

>150°C ↓ low pressure

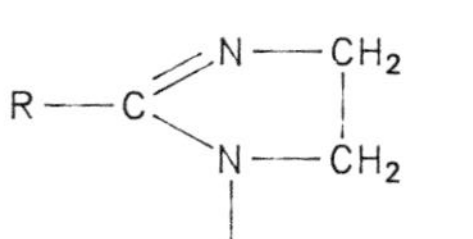

Equation 2

Figure 2 PREPARATION METHODS

In both reactions it is essential to complete the amidification stage before cyclisation commences. The fatty acid can be used in the form of an ester which in turn liberates the related alcohol during the first stage of the reaction as shown in Eq 3.

$$RCOOCH_3 + NH_2CH_2CH_2NHCH_2CH_2NH_2$$

$$\longrightarrow RCONHCH_2CH_2NHCH_2CH_2NH_2 + CH_3OH \qquad \underline{\text{Eq 3}}$$

These two products namely 2-alkyl 1-amino ethyl imidazoline and 2-alkyl 1-hydroxy ethyl imidazoline form the backbone of the vast array of imidazoline derivatives which have been developed during the past four decades. A selection of these imidazolines is covered later in this paper.

IV STRUCTURE OF 2-ALKYL IMIDAZOLINES

2-alkyl imidazoline has the following structure.

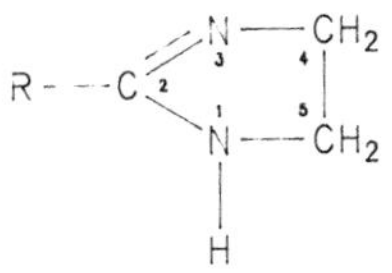

R—C(2)=N(3)—CH₂(4); C(2)—N(1)(H)—CH₂(5); CH₂(4)—CH₂(5)

where R is an alkyl radical generally containing 11 to 17 carbon atoms. The numbering system is shown with the No 1 position assigned to the nitrogen connected through single bonds to two carbon atoms of the ring and the No 3 position assigned to the nitrogen with a double bond connecting it with the No 2 carbon atom.

Further substitution at the No 1 position can give:

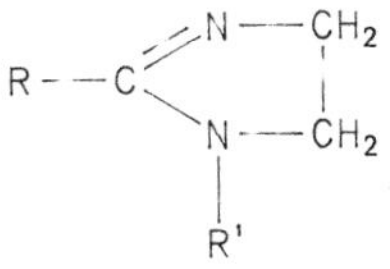

R—C=N—CH₂; C—N(R')—CH₂; CH₂—CH₂

where R is as above and R^1 can be any of the following:

— $CH_2\ CH_2\ NH_2$	-	amino ethyl
— $(CH_2\ CH_2\ NH)_n\ CH_2\ CH_2\ NH_2$	-	poly aminoethyl where n can be 1 to 5

— $CH_2\ CH_2\ OH$ - hydroxy ethyl

or other groups depending upon further reactions of R^1.

Although the name '2-alkyl imidazoline' is now the more common name, previously applied names included dihydroimidazole, dihydroglyoxaline and cyclic amidine. They readily form monohydrogen halide salts and titrate potentiometrically as mono acid bases.

As can be appreciated from the structure, 2-alkyl imidazolines contain both hydrophobic groups and hydrophilic groups which give the molecules their surfactant properties. This structure causes concentration of the surfactant at interfaces and hence a reduction in interfacial energy.
The molecule also orientates itself at the surface with the hydrophobic group directing away from the aqueous phase and its hydrophilic group actually in the aqueous phase. This surface phenomenon is shown in the diagram below.

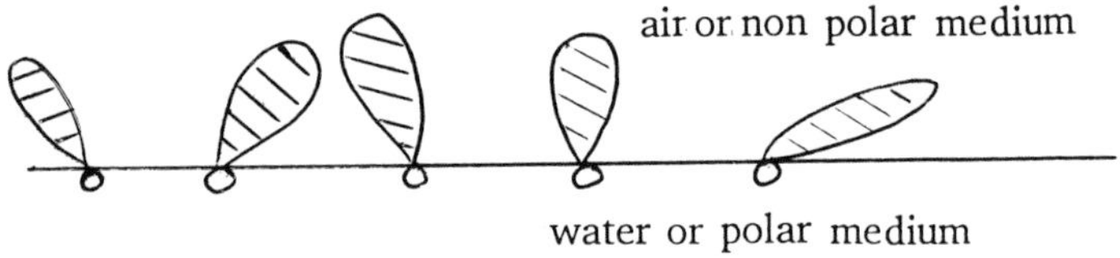

air or non polar medium

water or polar medium

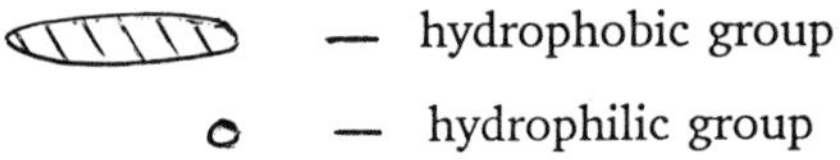

— hydrophobic group

— hydrophilic group

The imidazoline molecule is classed as a 'cationic' surfactant because there is a strong positive charge localised in one particular part of the molecule. This cationic nature leads to the molecule being strongly attracted to, and subsequently adsorbed by, negatively charged surfaces.

This specific property accounts for the majority of the interesting applications of alkyl imidazolines.

The balance between hydrophobic and hydrophilic properties depends on the chain length of R and the nature of R^1. In turn this can effect the detergency, emulsification, wetting and bactericidal properties. Examples of fatty acids which can be used in the preparation of 2-alkyl imidazolines are:-

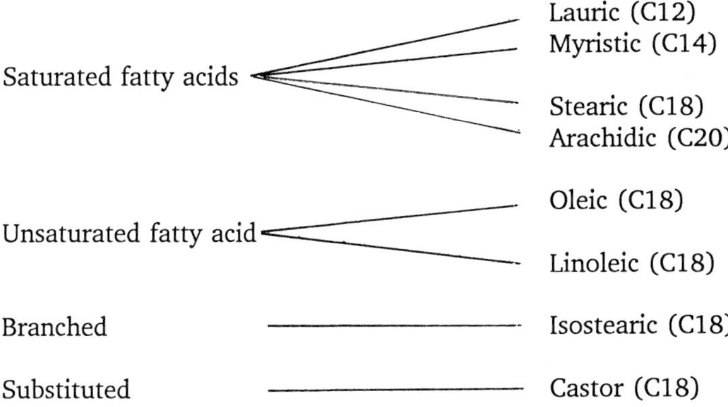

2-alkyl imidazolines with special properties can also be based on many other acids including phenoxyacetic acid, benzoic acid and naphthenic acid.

V REACTIONS OF 2-ALKYL IMIDAZOLINES

This section is split into four specific reaction areas, these are:

(a) Reactions of the amino ethyl group
(b) Reactions of unsubstituted imidazolines
(c) Reactions of the hydroxy ethyl group
(d) Reactions with alkylating agents

(a) Reactions of the amino ethyl group

These reactions are summarised in Figures 3 and 4 and some general comments are as follows. The ethylene oxide derivatives were first developed by Chwala in the late 1930's and were recommended for use in the textile industry[4]. The reaction products of 2-alkyl imidazolines and acrylic acid were patented in 1964 by Arndt and were recommended for use in acid and alkaline metal cleaners and also as emulsifiers in cosmetic formulations[5]. The mineral acid and fatty acid salts have been used extensively in corrosion inhibitor applications. Reaction with pentachlorophenol produced a salt which was found to be a very effective fungicide.

1-amino ethyl (or polyaminoethyl) 2-alkyl imidazolines will react with mono glycidyl compounds to give alkoxy hydroxy derivatives.

If the glycidyl group is difunctional then polymerisation can occur as shown in Eq 4.

$$2\,(-CH_2CH_2NH_2) + \overset{O}{\overbrace{CH_2-CH}} \sim R \sim \overset{O}{\overbrace{CH-CH_2}} \longrightarrow -CH_2-\underset{\underset{\underset{|}{CH_2}}{|}}{\underset{|}{\underset{CH_2}{N}}}-CH_2-CH(OH)\sim R \sim CH(OH)-CH_2-\underset{\underset{\underset{|}{CH_2}}{|}}{\underset{|}{\underset{CH_2}{N}}}-CH_2 \qquad \text{Eq 4}$$

Reaction with maleic anhydride yields imidazoline maleimides used as corrosion inhibitors[6]. Similarly reactions with phosphorylating agents yield imidazoline phosphoramidates which are used as corrosion inhibitors[7] .

* With Ethylene or Propylene Oxide

$$-CH_2CH_2NH_2 + 2\,\overset{O}{\overbrace{CH_2-CH_2}} \longrightarrow -CH_2CH_2N\langle^{CH_2CH_2OH}_{CH_2CH_2OH}$$

* With Acrylic Acid

$$-CH_2CH_2NH_2 + CH_2{=}CHCOOH \longrightarrow -CH_2CH_2NHCH_2CH_2COOH$$

* With a Mineral Acid

$$-CH_2CH_2NH_2 + HCl \longrightarrow -CH_2CH_2\overset{+}{N}H_3\overset{-}{Cl}$$

* With a Fatty Acid

$$-CH_2CH_2NH_2 + RCOOH \longrightarrow -CH_2CH_2NHCOR + H_2O$$

* With a Glycidyl Compound

$$-CH_2CH_2NH_2 + 2\,\overset{O}{\overbrace{CH_2-CH}}CH_2OR \longrightarrow -CH_2CH_2NHCH_2CH(OH)CH_2OR$$

* With Maleic Anhydride

$$-CH_2CH_2NH_2 + O\langle^{C(=O)-CH}_{C(=O)-CH} \longrightarrow -CH_2CH_2N\langle^{C(=O)-CH}_{C(=O)-CH} + H_2O$$

* With a Phenol

$$-CH_2CH_2NH_2 + HO-C_6Cl_5 \longrightarrow -CH_2CH_2\overset{+}{N}H_3O^{-}-C_6Cl_5$$

Figure 3 REACTIONS OF THE AMINO ETHYL GROUP

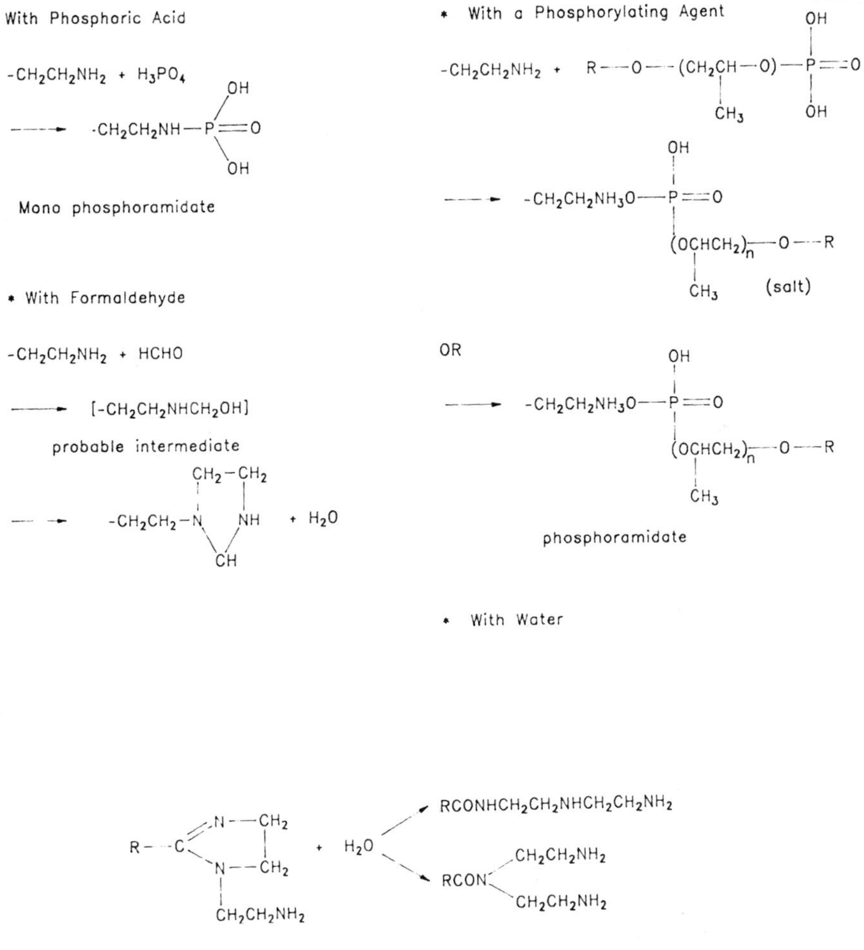

With Phosphoric Acid

$-CH_2CH_2NH_2 + H_3PO_4 \longrightarrow -CH_2CH_2NH-P(=O)(OH)_2$

Mono phosphoramidate

* With Formaldehyde

$-CH_2CH_2NH_2 + HCHO$

$\longrightarrow [-CH_2CH_2NHCH_2OH]$

probable intermediate

$\longrightarrow -CH_2CH_2-N(CH_2-CH_2-NH-CH) + H_2O$

* With a Phosphorylating Agent

$-CH_2CH_2NH_2 + R-O-(CH_2CH(CH_3)-O)-P(=O)(OH)_2$

$\longrightarrow -CH_2CH_2NH_3O-P(OH)(=O)-(OCH(CH_3)CH_2)_n-O-R$ (salt)

OR

$\longrightarrow -CH_2CH_2NH_3O-P(OH)(=O)-(OCH(CH_3)CH_2)_n-O-R$

phosphoramidate

* With Water

$R-C(=N-CH_2-CH_2-N(CH_2CH_2NH_2)) + H_2O \longrightarrow RCONHCH_2CH_2NHCH_2CH_2NH_2$

$\longrightarrow RCON(CH_2CH_2NH_2)_2$

Figure 4 REACTIONS OF THE AMINO ETHYL GROUP

Hydrolysis of amino ethyl alkyl imidazolines has been the subject of many discussions but evidence suggests that the hydrolysis yields a mixture of linear and tertiary amides as shown in Fig 4.

(b) Reactions of unsubstituted imidazolines

2-alkyl unsubstituted imidazolines show many of the reactions typical of secondary amines eg chlorohydroxy propane sultone yields products with excellent foaming, dispersing and wetting properties[8]. There are references in commercial literature to imidazoline metal complexes, however without detailed information of application.

(c) Reactions of the hydroxy ethyl group

These reactions are summarised in Figure 5 and some general comments are as follows:

The reaction with chloro sulphonic acid was developed by Johnson and Lee in 1961,[9] where the product was described as an 'ampholytic' composition and recommended for use in both textile and corrosion inhibitor applications. The reaction with the epichlorhydrin/bisulphite compound is attributed to Mannheimer in 1963,[10] who used his own previously patented 'sultone' as one of the reactants[11]. The interesting claim made for the product from reaction with propane sultone is compatibility with all classes of surfactants[12]. The products of reaction with acrylic acid depend upon whether the reaction takes place in anhydrous conditions or in the presence of water.

Evidence produced by Hein and Jaroschek supports the structures shown[13]. The products of reaction, often referred to as 'salt free' amphoterics, have found many uses especially in the textile industry as fabric softeners, and have also been used in metal cleaners and heavy duty spray degreasers.

Fig 6 shows one of the most important developments in imidazoline chemistry. Again Mannheimer was responsible for these developments in the 1950's and 1960's and in the original papers and patents, the possibility of hydrolysis of the imidazoline was not considered.

This particular point has been discussed at great length in previous papers and the reactions shown in Fig 6 are now generally accepted. Nevertheless, whichever structure, these products work well in a great variety of applications.

* With Chloro Sulphonic Acid

$$-CH_2CH_2OH + ClSO_3H \longrightarrow -CH_2CH_2{-}OSO_3H$$

* With Epichlorohydrin/Sodium Sulphite Compound

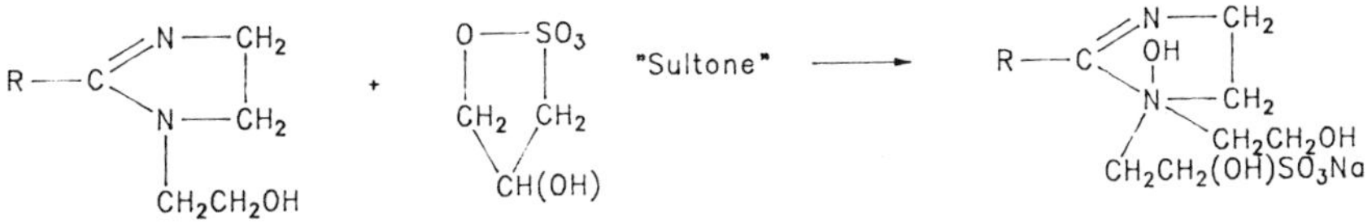

R—C(=N—CH₂)(N—CH₂)—N—CH₂CH₂OH + O—SO₃ / CH₂ CH₂ / CH(OH) "Sultone" ⟶ R—C(=N—CH₂)(OH—N—CH₂)(CH₂CH₂OH)(CH₂CH₂(OH)SO₃Na)

* With 1,3-Propane Sultone

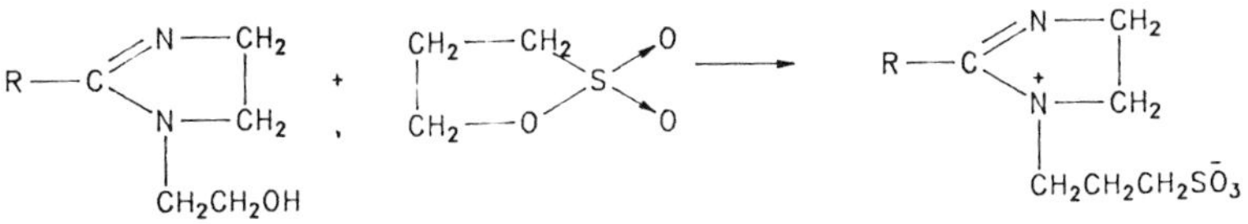

R—C(=N—CH₂)(N—CH₂)—N—CH₂CH₂OH + CH₂—CH₂—S(→O)(→O)—O—CH₂ ⟶ R—C(=N—CH₂)($\overset{+}{N}$—CH₂)—$CH_2CH_2CH_2S\bar{O}_3$

* With Acrylic Acid (Anhydrous)

$$-CH_2CH_2OH + CH_2{=}CHCOOH \longrightarrow -CH_2CH_2OCH_2CH_2COOH$$

* With Acrylic Acid (in presence of water)

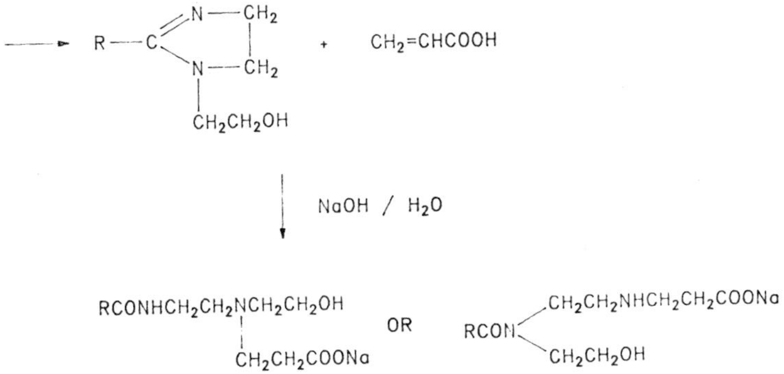

⟶ R—C(=N—CH₂)(N—CH₂)—N—CH₂CH₂OH + $CH_2{=}CHCOOH$

↓ NaOH / H_2O

$RCONHCH_2CH_2N(CH_2CH_2COONa)CH_2CH_2OH$ OR $RCON(CH_2CH_2NHCH_2CH_2COONa)(CH_2CH_2OH)$

Figure 5 REACTIONS OF THE HYDROXYL ETHYL GROUP

With Sodium Chloroacetate
(in presence of water)

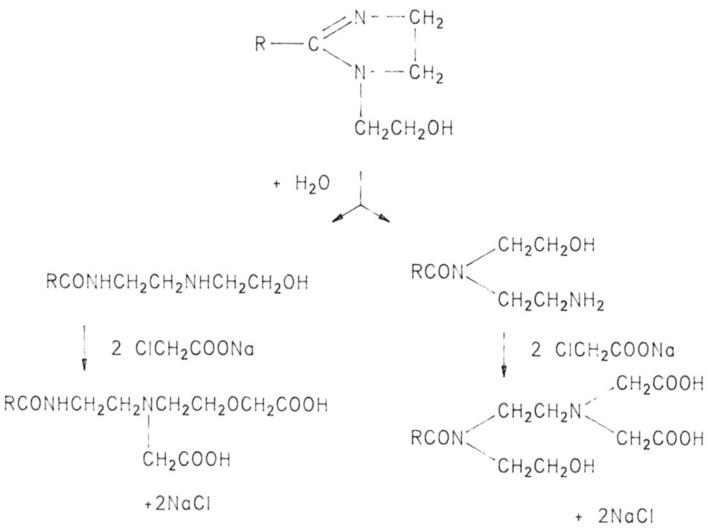

$$R-C\begin{cases}=N-CH_2\\ -N(CH_2CH_2OH)-CH_2\end{cases}$$

$+ H_2O$

$RCONHCH_2CH_2NHCH_2CH_2OH$ → ($2\ ClCH_2COONa$) → $RCONHCH_2CH_2N(CH_2COOH)CH_2CH_2OCH_2COOH + 2NaCl$

$RCON(CH_2CH_2OH)(CH_2CH_2NH_2)$ → ($2\ ClCH_2COONa$) → $RCON(CH_2CH_2OH)CH_2CH_2N(CH_2COOH)_2 + 2NaCl$

Figure 6 REACTIONS OF THE HYDROXYL ETHYL GROUP

They were named 'amphoterics' having cationic properties in the acid range and anionic properties in the alkaline range and of all the imidazoline based amphoterics are probably the most versatile ever developed.

The wide range of applications and continuing growth in use even today are very much related to the following properties:

a) Compatibility with other surfactant types
b) Low irritancy
c) Ability to lower the irritant effect of other products present in any mixture of surfactants.

Many excellent papers and chapters have been written on the subject of imidazoline based amphoterics[14].

(d) Reactions with alkylating agents

The general reaction scheme attributed to alkylation with XY is shown in Eq 5.

$$R-C(=N-CH_2-CH_2-N(CH_2CH_2OH)) + XY \longrightarrow \left[R-C(=N-CH_2-CH_2-\overset{+}{N}(X)(CH_2CH_2OH))\right] Y^-$$

Eq 5

Suggestions regarding the more likely structures for these compounds have been covered in previous work[15]. The earlier developments were used in bacteriostatic and bactericidal applications[16] but later were found to be useful deodorants, fungicides, algicides, anti-static agents and textile treatment products[17] . Eq 6 shows the formation of 1-(alkyl amido), 1-methyl, 2-alkyl imidazoline methosulphate used as a fabric conditioner.

$$R-C(=N-CH_2-CH_2-N(CH_2CH_2NHCOR)) + (CH_3)_2SO_4 \longrightarrow \left[R-C(=N-CH_2-CH_2-\overset{+}{N}(CH_3)(CH_2CH_2NHCOR))\right]$$

$CH_3SO_4^-$

Eq 6

VI APPLICATIONS

2-alkyl imidazolines find greatest use as cationic surfactants in fields which depend upon such properties as high surface activity, wettability, emulsifiability and biostatic or biocidal properties. Industrially, they have found uses in the textile, leather, paper, mining, coatings, agrochemicals, bitumen and corrosion inhibitor fields to name but a few.

This part of the discussion is in three sections covering some of these applications in relationship to the specific imidazoline type.

(a) Imidazolines and their salts
(b) Imidazoline quaternary ammonium compounds
(c) Imidazoline amphoterics

(a) Imidazolines and their salts

A list of the more well known applications of this class of compounds is shown below. Two of the applications, those of corrosion inhibition and epoxy resin curing will be subjected to a more detailed treatment.

Applications

Ore Flotation	CORROSION INHIBITION
Bitumen Emulsification	Bitumen Adhesion
EPOXY RESIN CURING	Textile Treatment
Pigment Flushing	Solvent Cleaners
Agricultural Sprays	Anti Static
Mildew Compositions	Pigment Wetting
Vaso Constrictors	Vaso Dilators

Tall oil fatty acid/diethylene triamine based imidazolines in the form of the acetate salts have been used extensively as ore flotation additives for many years.

2-alkyl imidazolines by themselves or as derivatives have been used as corrosion inhibitors in oil production since the 1940's. In nearly all cases they function as film forming inhibitors, consisting of a polar organic molecule, with a high concentration of positive charge localised in one part of the molecule and protect by forming a barrier at the cathodic sites.

Over the many years of use a variety of derivatives have been developed primarily to meet the broad range of corrosive environments encountered in the oil producing industry.

In most cases the parent imidazoline is based on a C18 fatty acid reacted with a polyethylene amine such as diethylene triamine (DETA) or triethylene tetramine (TETA). In order to meet the different solubility requirements which are encountered in use, the imidazolines are often converted to their mineral acid or fatty acid salts e.g.

Requirement		Modification
Oil sol/Water insol	-	Fatty acid salts
Oil sol/Water dispersibility	-	Imidazoline only
Oil insol/Water sol	-	Mineral acid salts
Oil sol/Water sol	-	Phosphoramidates

In the majority of applications the inhibitor is supplied in a diluted form and the usage rate is between 5-20 ppm.

As with many other chemicals the key requirements of an oil well corrosion inhibitor are low toxicity to marine life, biodegradability and performance, in that order. Whilst the parent imidazolines have always been considered to be of low toxicity, some of the derivatives including phosphoramidates are currently under examination.

The need for good corrosion inhibitors will continue and predicted growth rate for the USA is 4.5% over the next five years and 9% for Europe and the Middle East.

DETA based imidazolines with C17 alkyl chains were used as far back as the 1950's for emulsifying bitumen. A typical formulation would be as follows:

Bitumen	-	100 parts at 80 °C
Water	-	100 parts}
Imidazoline	-	1.5 parts} at 100 °C
Hydrochloric Acid	-	1.0 part}

In the late 1960's and onwards the fatty monamines and diamines were found to be superior to the alkyl imidazolines.

In the case of bitumen adhesion additives, the 2-alkyl imidazolines are still used primarily for their effectiveness and excellent thermal stability when stored in bitumen at high temperatures for long periods of time.

Again the ideal molecule is based on tall oil fatty acid (C18) and a polyethylene amine such as DETA, TETA or even higher amines to give improved thermal stability.

The use of poly amino ethyl alkyl imidazolines in epoxy resin technology is probably not initially regarded as a surfactant application since their main role is to provide reactive hydrogens to open the oxirane ring and effect crosslinking[18] . However some of the valuable properties imparted to an epoxy based coating by an alkyl imidazoline are certainly linked indirectly with their surfactant properties e.g excellent wettability of metal surfaces, excellent adhesion, good acid resistance and thermal resistance.

These particular epoxy curing agents were first developed by General Mills Inc in the 1950's and were based on the reaction between a dimerised fatty acid (C36) and a polyethylene amine such as DETA or TETA. Inevitably the reaction mixture contained imidazoline groups as shown below.

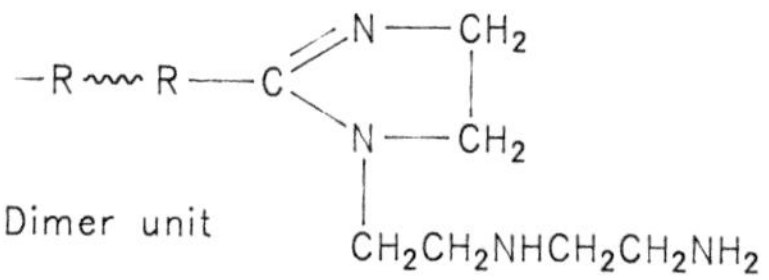

$$-R\sim\sim R-C\begin{matrix} \overset{N-CH_2}{//} \\ \underset{N-CH_2}{\backslash} \end{matrix}$$

Dimer unit $CH_2CH_2NHCH_2CH_2NH_2$

However the imidazoline molecule serves an additional function; that of the catalytic effect of the tertiary nitrogen on the self polymerisation of epoxy groups. This proceeds through the initial formation of an alkoxide ion as shown in Eqs 7 &8.

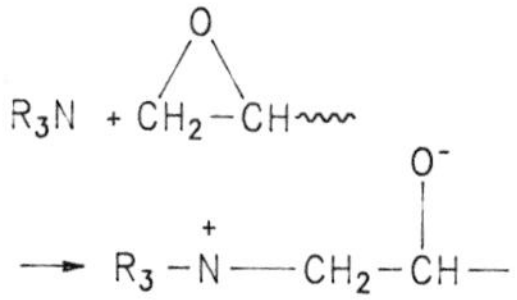

$$R_3N + \overset{O}{CH_2-CH}\sim\sim \longrightarrow R_3\overset{+}{N}-CH_2-\overset{O^-}{\overset{|}{CH}}- \qquad \text{Eq 7}$$

and then polymerisation

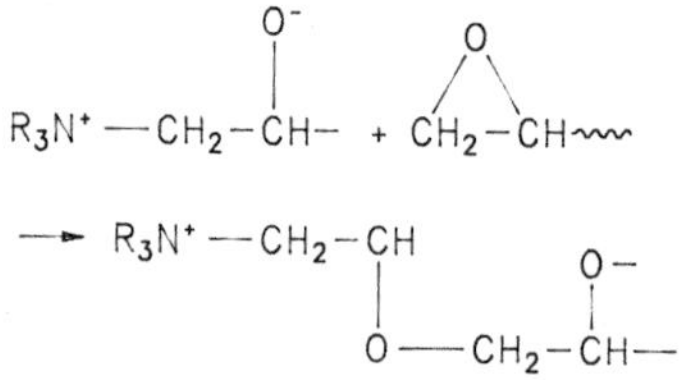

$$R_3N^+-CH_2-\overset{O^-}{\overset{|}{CH}}- + \overset{O}{CH_2-CH}\sim\sim \longrightarrow R_3N^+-CH_2-\underset{O-CH_2-\overset{O^-}{\overset{|}{CH}}-}{\underset{|}{CH}} \qquad \text{Eq 8}$$

and so on

During the 1960's the level of imidazoline present in the product was found to be very important and in some cases formed part of the specification. The so-called reactive polyamides generally contained approximately 60% of imidazoline relative to linear amide.

In addition to the improved coating properties there was a relationship between the imidazoline content and viscosity/cure rate.

In the areas of textile treatment, soil stabilisation, pigment flushing, solvent cleaners, pigment wetting and agricultural sprays the DETA based imidazolines with a C17 alkyl chain are still used in reasonable quantities.

Two interesting, but not necessarily surfactant linked applications of specific imidazolines are those of vasoconstrictors and vasodilators where the 2 positions are occupied by benzyl or benzyl amino substitutents[19].

(b) Imidazoline quaternary amononium compounds

These compounds have in the past played a very important part in the application of imidazolines and derivatives. They found many uses in the area of germicides, anti-static agents, fungicides, deodorants, algicides and fabric softeners.

Fungicidal activity, like antibacterial activity tends to be specific with regard to the test organism. In general, quaternary cationic surfactants including imidazoline quaternaries, exhibit a wide fungicidal spectrum[20].
Shephard and Shonle considered a variety of quaternary imidazoline compounds to be very effective as topical antiseptics[21].

A quaternary 1-hydroxy ethyl imidazoline based on ethyl bromide has proved to be very effective as an algicide and is particularly useful in the control of algae in swimming pools. Deodorants for use in breweries, fish and meat processing plants can be formulated using the above mentioned ethyl bromide quaternary.

Excellent fabric softeners can be produced by quaternising an alkyl amido alkyl imidazoline with dimethyl sulphate as shown in Eq 6. The need for fabric softeners arose from the change from traditional soap to the use of detergents which latter gave fabrics a ‘harsh’ feel. Cationic based fabric softeners for domestic use were first developed in the USA and then spread into Europe via Germany. Today it is an enormous business and still growing. There are three chemical types used:

Type 1

dialkyl dimethyl ammonium chloride

$$\left[R_2 N^+ (CH_3)_2 \right] Cl^-$$

Type 2

2-alkyl, 1-methyl, 1-(alkyl amido ethyl) imidazoline metho sulphate

$$\left[R{-}C{\overset{\displaystyle =N-CH_2}{\underset{\displaystyle N^+(CH_3)(CH_2CH_2NHCOR)-CH_2}{}}} \right] CH_3SO_4^-$$

Type 3

diamido alkoxy methyl metho sulphate

$$\left[RCONHCH_2CH_2\overset{\displaystyle CH_3}{\underset{\displaystyle (CH_2CH_2O)_nH}{N^+}}CH_2CH_2NHCOR \right] CH_3SO_4^-$$

Normally these products are offered as 75% concentrates in iso propanol. Fabric softeners perform three functions: softening, anti-static and re-wetting. Type 1 softener has generally been accepted as giving the best overall softening properties[22], with Type 2 not far behind. Specific modification of Type 2 can bring the softening properties to be equal or better than Type 1.

Many fabrics, especially those made from synthetic fibres, possess a surface charge referred to as 'static charge', which can impart undesirable properties to the fabric e.g. the fabric tends to stick together.

Quaternary ammonium compounds are well known for their ability to remove this static charge and are often referred to as 'anti-static' agents. Therefore it is not surprising that all of the fabric softener types described impart anti-static properties, in addition to their softening action. The imidazoline quaternary (Type 2) has been rated as the best product to use for maximum anti-static effect[23] .

Rewetting properties are very important since excessive use of a specific type of softener can ultimately waterproof the fabric. Of the imidazoline quaternaries, Type 2 is considered better than Type 1 with respect to re-wetting properties[24] .

Therefore, although Type 1 has been the most widely used in many formulations, blends of Type 1 and Type 2 have been used to give a balance of the properties described above. Because of viscosity characteristics imidazoline quaternaries aid the production of 'concentrated' formulations and, additionally this type is thought to be more acceptable with regard to biodegradability than Type 1.

The key features of an imidazoline quaternary used in fabric softening are:-

- Di alkyl substitution
- C17 chain length
- Straight chain rather than a branched chain
- Saturated alkyl chains

The use of fabric softeners will continue to grow; the only change likely to be in the manner of application. In the past, most fabric softeners have been applied at the rinse cycle stage rather than the wash or dryer stages. Future developments will be directed at improving the fabric softeners for use in the wash and dryer stages.

(c) Imidazoline based amphoterics

Like the previous two imidazoline types, the amphoterics have been used successfully across a broad range of applications as shown below.

Applications

Shampoos	Body Cleansers
Body Lotions	Conditioners
Fabric Detergents	Fabric Softeners
Anti Static Agents	Corrosion Inhibition
Metal Cleaners	Bitumen Emulsion Breaking
Pigment Dispersants	Eye Make Up Remover
Bath Oils	

The term amphoteric was first introduced into surfactant chemistry by Mannheimer and he and the Miranol Corp went on to fully exploit the rather unique properties of these compounds.

There has never been any argument about the main reason behind the early growth of these products, which was their very low level of irritancy compared with many other surfactant types. This feature was used to impart mildness to hair shampoos based on these amphoterics. Furthermore they tended overall to reduce the irritant properties of co surfactants.

As previously described two of the more important classes of imidazoline based amphoterics are formed by reacting the imidazoline with sodium mono-chloroacetate or acrylic acid. In the first case, the reaction has to be conducted in the presence of water and it is now generally accepted that hydrolysis occurs resulting in the compounds shown in Fig 5.

With acrylic acid, the resultant product depends on whether the reaction is conducted in anhydrous conditions or not (see Fig 4) Apart from the low irritating properties the amphoterics possess stability to extreme pH levels, tolerance to hard water, inbuilt corrosion inhibition, bactericidal properties, biodegradability, substantivity and tolerance to high levels of electrolytes.

Many excellent papers and books have been devoted to the subject of amphoterics in general, including imidazoline based amphoterics.

Reference should be made to such books and articles for detailed information regarding specific application formulations[25]. Formulations can contain upto 10% of the amphoteric and depending upon the application, may contain colorants, perfumes, preservatives, thickeners, foam boosters, other surfactants and pH adjusters (eg citric acid).

European Surfactant Usage

The following table gives a summary of the current usage of surfactants across Europe.

Table 1 (All figures in tonnes)

Anionics	-	2, 730, 000
Nonionics	-	920, 000
Imidazoline based cationics	-	99, 000
Other cationics	-	221, 000
Amphoterics	-	22, 000
Others	-	2, 000
Total	-	3, 994, 00

As the above figures show the imidazolines and derivatives still represent only a fraction of the total surfactant industry, in fact, a mere 2.5%. Nevertheless their role has been, and still is, quite unique and very important. In the majority of cases specific performances cannot be matched by their close relatives, the monamines and diamines.

The author is confident that the imidazolines and their derivatives have an assured and growing future and, because of their low toxicity properties, may begin to replace other cationics in applications where toxicity to marine life and biodegradability are in question.

REFERENCES

1 A.W. Hofmann, Ber. 21, 1888, 2332

2 E.Waldmann U.S. Pat 3,711,404 (1973); A.Chwala, U.S. Pat 2,211,001 (1940); B.G. Wilkes, A.L. Wilson, U.S. Pat 2, 268, 273 (1941); P.F. Tryon, U.S. Pat 2.520, 102 (1950), H.Shonle, E. Shepard, U.S. Pat 2, 493, 318, 2, 493, 320, 2, 493, 321, 2, 493, 322 (1950); W.B. Hughes, U.S. Pat 2, 846, 440 (1958); D.Redmore, U.S. Pat 3, 711, 404 (1973); H. Mannheimer, U.S. Pat 2, 773, 068 (1956); C.E. Johnson, U.S. Pat 2, 985, 662 (1961); W.B. Hughes, U.S Pat 2, 918, 474 (1959).

3 R.J. Ferm, J.L. Riebsomer, Chem. Rev, 1954, 54(4), 593.

4 A. Chwala, U.S. Pat 2, 211, 001 (1940).

5 G.J. Arndt, B. Pat 1, 078, 101 (1964).

6 W.B. Hughes, U.S. Pat 2, 918, 474 (1959).

7 D. Redmore, U.S. Pat 3, 711, 404 (1973).

8 E. Waldmann, U.S. Pat 2, 154, 922 (1939).

9 C.E. Johnson, F.G. Lee, U.S. Pat 2, 985, 662 (1961).

10 H.S. Mannheimer, U.S. Pat 3,231, 580 (1966).

11 H.S. Mannheimer, U.S. Pat 3, 100, 779 (1963).

12 R. Ernst, U.S. Pat 3, 280, 179 (1966).

13 H. Hein, H.J.Jaroschek and W. Melloh, Cosmetics and Toiletries, 1980, 95, 37.

14 B.R. Bluestein, C.L. Hilton 'Amphoteric Surfactants', Marcel Dekker, NY, 1982, 12, 1, 1; E. Lomax Cosmetics and Toiletries, Manufactureres and Suppliers, 1991, Issue XX1, 20; H. Hein, H. J. Jaroschek and W. Melloh, Cosmetics and Toiletries, 1980, 95, 45; E Lomax, Manufacturing Chemist and Aerosol News, 1980, 1979, 39; D. Bass, Soap, Perfumery and Cosmetics, 1977.

15 B.R. Bluestein, C.L.Hilton, 'Amphoteric Surfactants', Marcel Dekker, NY, 1982, 12, 1, 6.

16 E.R. Shepard, H.A. Shonle, J. Am. Chem Soc, 69, 2269; E.R. Shepard, H.A. Shonle, U.S. Pat 2, 493, 322 (1950); E.R. Shepard, H.A. Shonle, U.S. Pat 2, 493, 320 (1950).

17 B.G. Wilkes, A.L. Wilson U.S. Pat 2, 268, 273 (1941).

18 H.Lee, K.Neville, 'Handbook of Epoxy Resins', McGraw-Hill, NY, 1967, 7, 1.

19 R.J. Ferm, J. Reibsomer, Chem Rev, 1954, 54 (4), 593.

20 A. Schwartz, J. Perry, J. Berch, 'Surface Active Agents and Detergents', Interscience, NY, 1958, 11, 7, 239.

21 A. Schwartz, J. Perry, J. Berch, 'Surface Active Agents and Detergents', Interscience, NY, 1958, 11, 7, 218.

22 R.R. Egan, J. Am. Oil Chemists Soc., 1978, 55, 120.

23 R.R. Egan, J. Am. Oil Chemists Soc., 1978, 55, 120.

24 R.R. Egan, J. Am. Oil Chemists Soc., 1978, 55, 120.

25 B.R. Bluestein, C.L. Hilton, 'Amphoteric Surfactants', Marcel Dekker, New York, 1982, 12, 1, 6; E. Lomax, Cosmetics and Toiletries, Manufacturers and Suppliers, 1991, Issue XX1, 20.

Betaine Esters: Quaternary Ammonium Compounds With Time-limited Activity

L. Edebo, B. Ahlström, S. Allenmark, M. Bertilsson, E. Jennische, S. Lange, M. Lindstedt, R.A. Thompson

DEPARTMENT OF CLINICAL BACTERIOLOGY, UNIVERSITY OF GÖTEBORG, S-413 46 GÖTEBERG, SWEDEN

1 INTRODUCTION

Most antimicrobial agents and drugs remain active after their objective has been performed which may cause unwanted side-reactions. By designing substances which disintegrate into inactive products, such side-reactions may be avoided. Such compounds have been called soft drugs[1]. The antimicrobial effect of membrane-active, amphiphilic quaternary ammonium compounds is extremely rapid implying that usually they remain long after the antimicrobial action has been performed. However, quaternary ammonium compounds can be made up of esters of betaine with fatty alcohols. In addition to exerting antimicrobial activity, such compounds decompose spontaneously into substances which are non-toxic and readily metabolized. We have been studying the mechanisms of antimicrobial activity and hydrolysis in order to find optimum conditions for application in time-limited antisepsis, disinfection and extraction of cellular components such as nucleic acids and enzymes. Both the hydrolysis and the antimicrobial effect are influenced - often extensively - by pH, temperature, presence of salts, fatty acids and proteins.

These effects, as well as the release of the periplasmic enzyme β-lactamase from *E. coli*, effects on the mammalian mucosa, surface disinfection of chicken, and inactivation of sexually transmitted agents and other genital microbes will be discussed.

2 CHEMICAL MECHANISMS

Besides their well-known antimicrobial effect, cationic amphiphiles show many other interesting properties, mainly associated with their tendency to self-aggregate in solution. The micellar systems that can be created by quaternary ammonium compounds like cetyltrimethylammonium bromide (CTAB) and analogues, have been extensively studied and utilized for a variety of different purposes. However, less has been done with respect to chemical modification of the stable cationic amphiphiles of this type in order to make them easily degradable into non-surface-active components - a property which could be useful for many practical applications. In the following, we describe how incorporation of a hydrolysable structure element can lead to interesting and useful chemical and biological behaviour.

Structure and General Chemical Properties

The general structure of the class of compounds we have investigated is given in Fig. 1. This structure incorporates esters of amino acids (R = H), as well as the betaine-derived esters (R' = CH_3, R'' = H) bearing a permanent positive charge. The latter compounds, being the subject of this review, have been synthesised by reaction of chloroacetyl chloride with a series of alkanols (C_4-C_{18}), followed by quaternisation with triethylamine[2]. Radiolabelling of these compounds has also been effected using the same main synthetic route. A doubly labelled compound[3] is achievable via ^{14}C-bromoacetic acid and a ^{3}H-labelled alkanol (easily obtained by lithium aluminium hydride reduction of the corresponding tritiated alkanoic acid precursor).

From the structure given in Fig. 1, it is clear that a certain charge repulsion is present in the molecular ground state (due to the dipolar character of the ester carbonyl group). Therefore, all reactions involving attack of an anionic nucleophile at the carbonyl carbon atom in the rate-determining step will be faster than in normal esters. Conversely, reactions involving protonation of the carbonyl oxygen atom will increase the charge repulsion

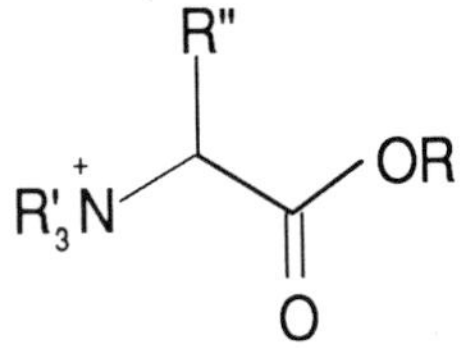

R''
$R'_3\overset{+}{N}$
OR
O

Figure 1

and consequently take place at a much slower rate. Since ester hydrolysis can proceed via both acid- and base-catalysed mechanisms, it is evident that in this special case the first mechanism is extremely disfavoured, whereas the opposite holds for the second mechanism. This gives rise to an unusually large pH-dependence of the hydrolysis of such alkyl betainates. The dependence of the short-chain betainates was studied already in the 1950's by Robson Wright[4,5]. We have found that this pH-dependence also holds true for the long-chain esters[2] (Fig. 2). It is noteworthy that large effects on the rate occur in a pH-region around the physiological pH.

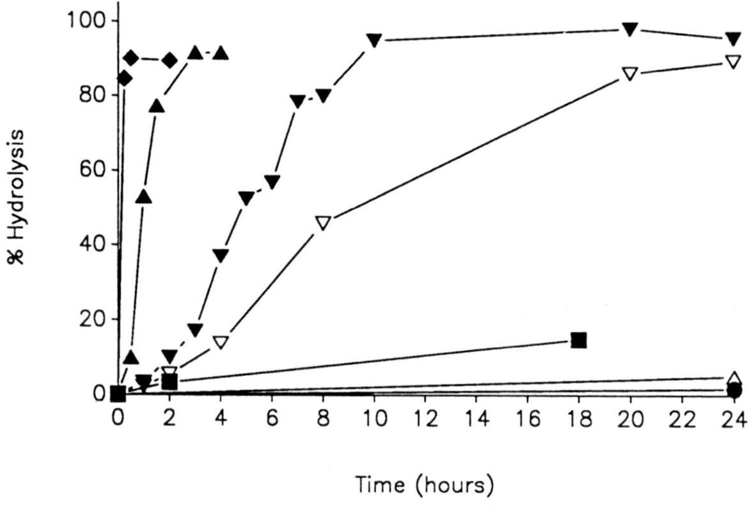

Figure 2. Dependence on pH of the rate of hydrolysis of tetradecyl betainate at pH 3 (●), pH 5 (Δ), pH 6 (■), pH 7 (▼), pH 8 (▲), and pH 9 (◆), all at 30°C, except that the hydrolysis rate at pH 7 was also tested at 25°C (∇). (Reprinted, with permission, from M. Lindstedt, S. Allenmark, R. A. Thompson and L. Edebo, *Antimicrob. Agents Chemother.* **34**, 1949 (1990). Copyright 1990, American Society for Microbiology.)

Effects Arising from the Amphiphilic Nature of Long-Chain Esters

Studies of the influence of the alkyl chain length upon the rate of hydrolysis of compounds, where one of the R' groups = p-nitrophenyl, revealed that a more complex mechanism was operating in the case of the amphiphilic esters (*i.e.* for chain lengths > ca C_{10})[6]. This abnormal behaviour is expressed as a very pronounced concentration dependence of the observed rate constant, leading to a substantial rate increase at concentrations close to the cmc. The situation is illustrated in Fig. 3.

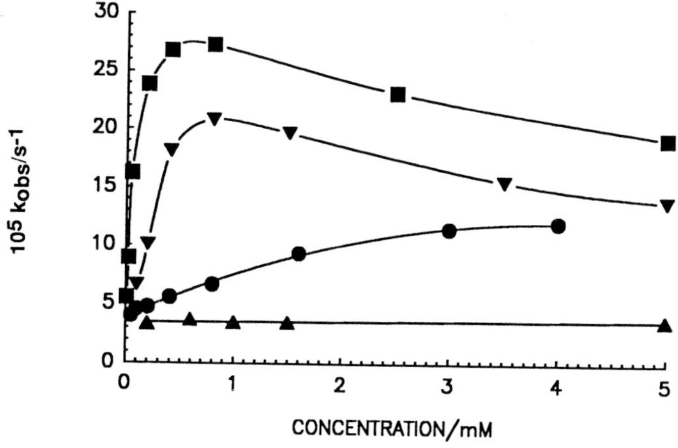

Figure 3. Effect of concentration and chain length on the rate of hydrolysis. R = C_3 (▲), C_{10} (●), C_{12} (▼), and C_{14} (■). (Reprinted, with permission, from R. A. Thompson and S. Allenmark, *Acta Chem. Scand.* **43**, 690 (1989).

The rate of hydrolysis increases with increasing ester concentration and reaches a maximum close to the cmc. Further, the maximum rate increase is dependent of the length of the alkyl chain in the alkanol part of the ester, the longer the chain length the greater the rate increase. The catalysis is a result of an increased concentration of the reactants in the micellar Stern layer, which functions as a pseudophase[7-10]. The hydroxide ion concentration in the Stern layer is partially a result of the amphiphile and the resulting micelle, but also a result of the medium, especially the concentration of other counter ions. It is expected that there will exist an ion-exchange equilibrium between ions in the micellar and water phases. The micellar catalysis studied here is of a rather special type, however, since the reacting substrate is the same as the micelle-forming agent.

Since the amphiphilic compounds undergoing micellization, and thereby self-induced catalysis, were also active as antimicrobial agents, we synthesized a second series of compounds without the chromophore (R" = H, R' = CH_3), but radiolabelled with ^3H or ^{14}C, or dual labelled with both isotopes. The hydrolysis rates were then followed by HPLC with an on-line radioisotope detector. These compounds were used to obtain the exact hydrolysis rates for the compounds of biological interest[11].

The cmc of the tetradecyl betainate is 1.32 mM in 10 mM phosphate buffer as measured by conductivity. As can be seen from Fig. 4 there is a considerable catalytic effect even below the cmc. We attribute this effect to the presence of submicellar aggregates. These aggregates could cause the catalytic effect either through the creation of a microenvironment that increases the reactivity of the the ions with the ester or through an increased concentration of hydroxide ions.

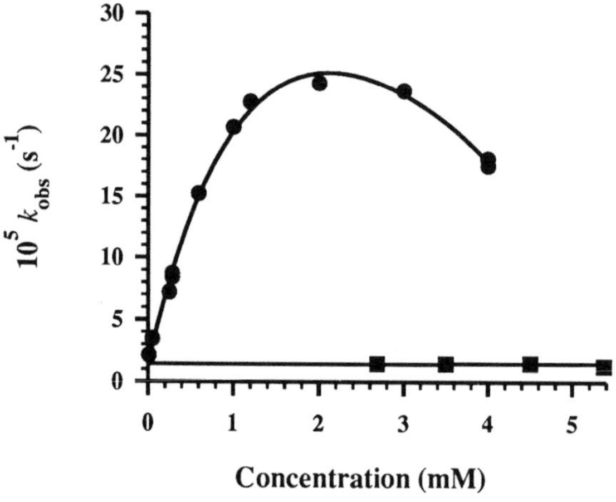

Figure 4. Dependence of rate constant on ester concentration. R = C_3 (■) and C_{14} (●). (Reprinted, with permission, from R.A. Thompson and S. Allenmark, *J. Colloid Interface Sci.*, (1992) in press. Copyright 1992, Academic Press, Inc.)

Factors Influencing the Rate of Hydrolysis.

There is a large temperature effect on the hydrolysis rate. An Arrhenius plot from kinetic experiment carried out at pH 7.5 (10 mM phosphate buffer) is shown in Fig. 5. The enthalpy and entropy of activation found were $\Delta H^{\ddagger} = 91.1$ kJ mole^{-1} and $\Delta S^{\ddagger} = 114$ J K^{-1} mole^{-1}, respectively.

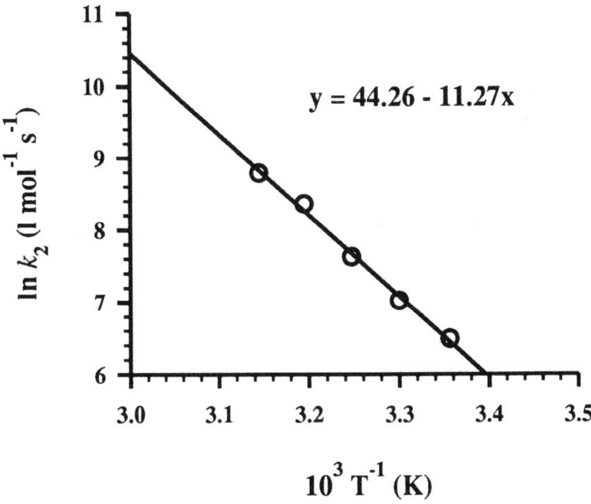

Figure 5. Effect of temperature on the rate of hydrolysis of tetradecyl betainate. (Reprinted, with permission, from R.A. Thompson and S. Allenmark, *J. Colloid Interface Sci.*, (1992) in press. Copyright 1992, Academic Press, Inc.)

Although a slight negative salt effect has been observed on the hydrolysis of methyl betainate[5], the effect is much larger for the amphiphilic compounds undergoing micellar catalysis[11]. The rate retarding effect increases on going from Cl^- to Br^- (Fig. 6). Our results are well in agreement with the pseudophase ion-exchange model, meaning that the local concentration of OH^- in the micelle's Stern layer is decreased due to competition by other anions such as halides. Since this local OH^- concentration (which is much higher than in the bulk of the solution) is responsible for the increased rate of hydrolysis, any displacement by competing anions will lead to a diminished rate. The pseudophase ion-exchange model describes the concentration of the different ions in the two phases in terms of an ion-exchange mechanism. It has been shown earlier that large low density ions such as Cl^- and Br^- displace small high charge density ions such as OH^- from the micellar phase[8,12,13].

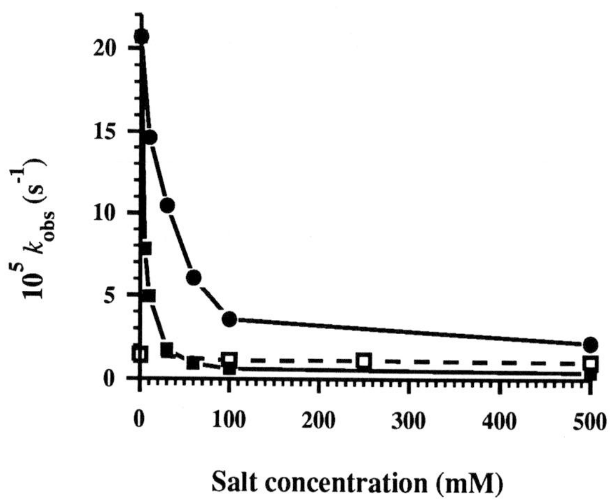

Figure 6. Effect of hydrophilic ions on the rates of hydrolysis. R = C_3: effect of NaBr (□). R = C_{14}: effect of NaBr (■) and NaCl (●).

We have also studied the effect of hydrophobic ions on the rates of hydrolysis[11]. Figure 7 shows the effect of hexanoic and tetradecanoic acid on the hydrolysis rate of tetradecyl betainate (B-14). As can be seen, the decrease in the hydrolysis rate is much more drastic than in the case with the hydrophilic ions (Br^- and Cl^-). This is as would be expected, since hydrophobic anions will interact with cationic micelles through both electrostatic and hydrophobic interaction. As can be seen in Fig. 7, tetradecanoic acid not only completely suppresses the micellar catalysis of the hydrolysis, but also more or less suppresses even the non-micellar catalysed hydrolysis. This is probably due to the formation of mixed micelles which inhibit the incorporation of hydroxide ions into the micelle. These micelles will at the same time lower the cmc of the system and, since tetradecanoic acid is in excess, probably almost completely deplete the aqueous phase of free betaine ester.

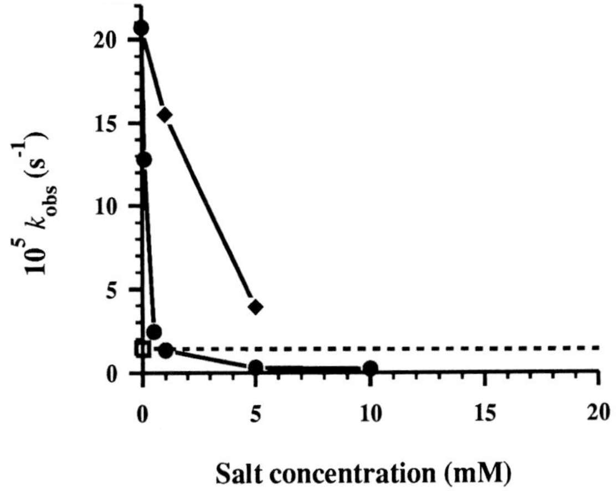

Figure 7. Effect of hydrophobic ions on the rate of hydrolysis. R = C_3: effect of NaBr (❑) (as reference). R = C_{14}: effect of hexanoic acid (◆) and tetradecanoic acid (●).

3 ANTIMICROBIAL MECHANISMS

Physicochemical Effects.

Initially[2], the antimicrobial effect of betaine esters was compared to that of stable quaternary ammonium compounds by measuring the inactivation of *Salmonella typhimurium* (smooth), *Pseudomonas aeruginosa, Bacillus megaterium* and *Candida albicans* by B-14 as well as by cetyltrimethylammonium bromide (CTAB). It was found that 2 μM concentrations were sufficient to attain >3.5 log reduction of the viable counts of *B. megaterium* in 10 min. whereas 10 μM was required for *S. typhimurium* and *P. aeruginosa* (Fig. 8). In *C. albicans* 50 μM of B-14 for 30 min. produced ca -3 log inactivation. The inactivation kinetics were similar, but the rate was slightly lower with B-14 than with CTAB. The similarities in chemical structure and inactivation kinetics, as well as the greater sensitivity of *B. megaterium* (gram-positive) than *S. typhimurium* and *P. aeruginosa* (gram-negative) and the still lower sensitivity of *C. albicans* (yeast), are evidence that betaine esters show the same mechanism of action as the stable quaternary ammonium compounds[14]. This conclusion has not been contradicted by any results achieved so far.

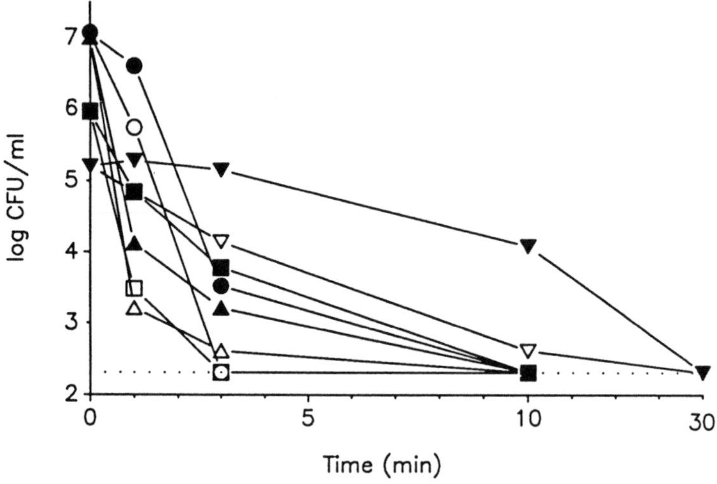

Figure 8. Antimicrobial activities of tetradecyl betainate (B-14) (filled symbols) and CTAB (open symbols) in 10 mM sodium citrate buffer (pH 6.0) at 30°C. Against *S. typhimurium* 395MS (● and ○) and P. aeruginosa (▲ and △), 10 µM concentrations of the compounds were used; against *B. megaterium* (■ and □), a 2 µM concentration was used; and against C. *albicans* (▼ and ▽), a 50 µM concentration was used. Corresponding effects were observed at other concentrations (data not shown). The limit of detection of the procedure is indicated by the dotted line. (Reprinted with permission, see Fig. 2).

Effect of hydrocarbon chain length. The susceptibility of suspensions of *S. typhimurium* in dilute buffer solutions to betaine esters increased with the length of the hydrocarbon chain to at least 16-18 atoms (Fig. 9). Like stable quaternary ammonium compounds the betaine esters probably interact with the lipid bilayer structures of the outer and the plasma membranes[15-17]. The phospholipids of both types of membranes contain fatty acids, mainly C_{16} and C_{18}[18]. Furthermore, the lipopolysaccharide (LPS) of *Salmonella* contains C_{12} and C_{16} saturated fatty acids esterified to the 3-hydroxytetradecanoic acid residues which are amide linked to the glucosamine moieties of lipid A[19]. This structure will accommodate a hydrocarbon chain length in the outer membrane of at least 18 carbon atoms. Thus, the high microbicidal effect of the C_{16} and C_{18} betaine ester probably results from good fit and extensive hydrophobic interaction between the hydrocarbon chains of the bacterial membranes and those of the betaine esters in combination with charge interaction between the quaternary nitrogen and negative membrane charges[20].

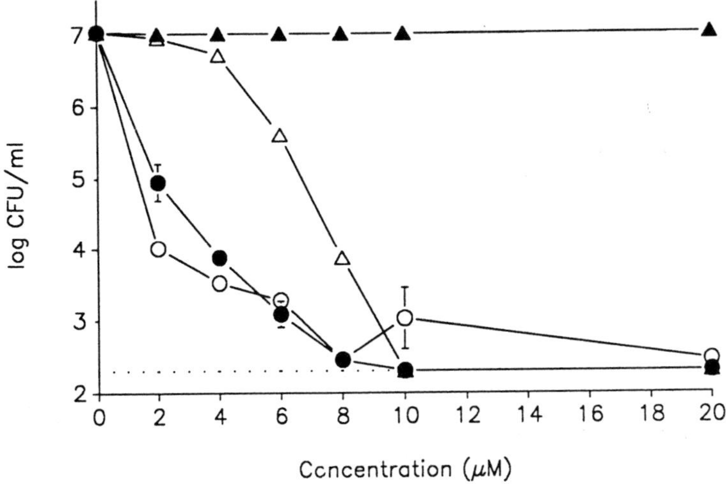

Figure 9. Relationship between length of the alkyl chain and bactericidal activity of betaine esters against *S. typhimurium* 395MS in 10 mM sodium phosphate buffer (pH 7.0) at 30°C. The length of contact was 3 min. Each point represents the mean of two separate experiments, and the duplicate values are connected by a vertical bar, most often covered by the symbol. The limit of detection of the procedure is indicated by the dotted line. Symbols: ▲, dodecyl betainate (B-12); Δ, tetradecyl betainate (B-14); ●, hexadecyl betainate (B-16); O, octadecyl betainate (B-18). Decyl betainate (B-10) showed the same result as B-12. (Reprinted with permission, see Fig. 2).

Effect of pH. The increase of the bactericidal effect at higher pH (Fig. 10) may, in consequence, result from stronger charge interaction due to greater negative charge of the bacterial membranes, as noted before with stable compounds[14].

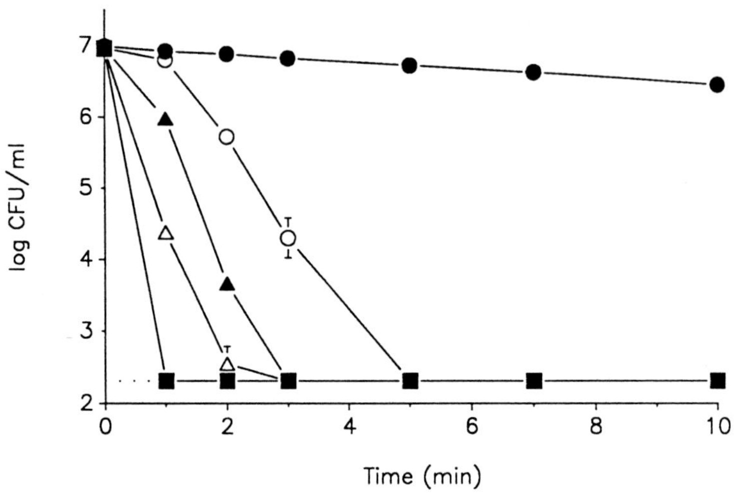

Figure 10. Effect of pH on the bactericidal activity of 5 µg of tetradecyl betainate (B-14) per ml (14 µM) at 30°C against *S. typhimurium* 395MS. Each point represents the mean of two separate experiments, and the duplicate values are connected by a vertical bar, most often covered by the symbol. The limit of detection of the procedure is indicated by the dotted line. Symbols: ●, pH 4.0; ○, pH 5.0; ▲, pH 6.0; Δ, pH 7.0; ■, pH 8.0. At pH 9.0 as well as at pH 8.0 complete killing was achieved in 1 min. (Reprinted with permission, see Fig. 2).

Influence of protein on optimum of pH and hydrocarbon chain length. A different ranking emerged, when the suspension medium contained 0.2% bovine serum albumin (BSA). B-14 was more efficient than B-16 (Table 1) probably due to less binding to BSA. This may also be the reason for the greater antimicrobial activity at pH 6 than at pH 7 in the presence of BSA, which was a reversal of the effect of pH in the absence of BSA. In the presence of 1% BSA the reversed effect of pH was more pronounced (Fig. 11). At pH 9 the killing of *S. typhimurium* was inconspicuous, in this case B-14 was dissolved in ethanol and then added to bacteria suspended in the BSA solution to prevent hydrolysis of B-14 prior to the addition of bacteria. Obviously there was a very fast binding of B-14 to albumin which prevented its bactericidal action. To what extent the binding to BSA affected the liability of the betaine esters to hydrolysis at high hydroxide concentrations is presently unknown.

Table 1. Inactivation of *S. typhimurium* 395MS, *P. aeruginosa* 10783 and *C. jejuni* C5 Hag by 30 µg/ml tetradecyl (B-14) or hexadecyl betainate (B-16) in the presence of 0.2% BSA at 30°C for 10 min.

Buffer	pH	*S. typhimurium* B-14	*S. typhimurium* B-16	*P. aeruginosa* B-14	*P. aeruginosa* B-16	*C. jejuni* B-14	*C. jejuni* B-16
citrate	6	>-3.7[a]	-3.7	>-3.7	-3.7	-3.7	- 3.7
phosphate	6	3.2	3.1	n.d.	n.d.	n.d.	n.d.
citrate	7	>-3.7	-2.8	-3.7	-1.9	-3.7	-3.7
phosphate	7	-1.8	-0.9	>-3.7	-1.6	>-3.7	-3.4
tris	7	-3.4	>-3.7	-1.2	-1.1	>-3.7	>-3.7

[a]Initial conc. 10^5 cfu/ml. Lowest measurable conc. 20 cfu/ml (log 1.3)
n.d. = not determined

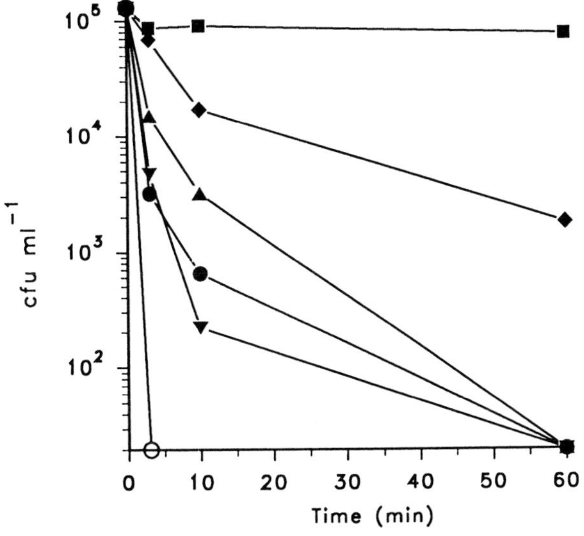

Figure 11. Effect of pH on the bactericidal activity of 100 µg tetradecyl betainate (B-14) per ml against *S. typhimurium* 395MS in 0.1 M buffers with 1% bovine serum albumin (BSA). Symbols: ○ nutrient broth, ▼ phosphate pH 6.0; ▲ phosphate pH 7.0; ● tris pH 7.0; ◆ tris pH 8.0, and ■ tris pH 9.0.

Competitive inhibition by protein. The inhibitory effect of BSA to different concentrations of B-14 at pH 6 and 30°C (Fig. 12) showed characteristics of competitive inhibition with respect to killing of *S. typhimurium*. The concentrations of B-14 required to achieve a certain bactericidal effect increased with the concentration of BSA, the relationship between B-14 and BSA being approximately constant. In order to achieve 99.99% (-4 log) inactivation in 10 min, B-14 concentrations 104 times less (SD = ± 30% in a rather coarse titration) than those of BSA sufficed on a weight basis. On a molar basis, this corresponds to 2.1 B-14 molecules per BSA molecule.

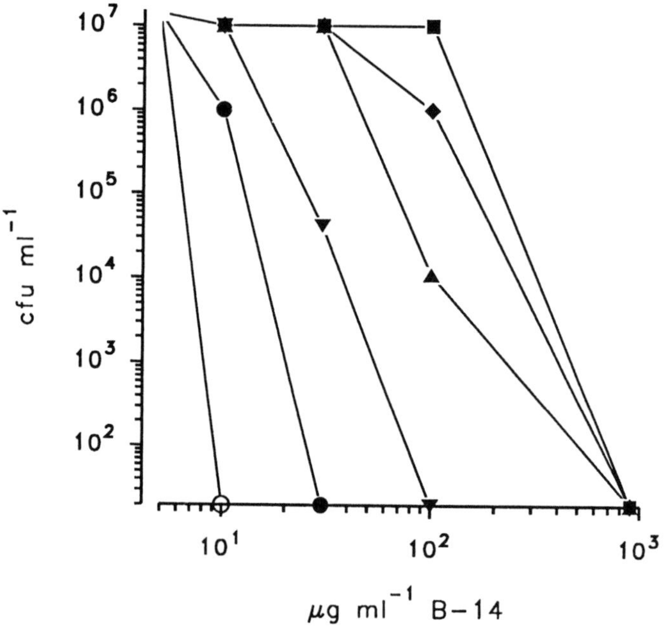

Figure 12. Effect of bovine serum albumin (BSA) on the bactericidal effect of tetradecyl betainate (B-14) against *S. typhimurium* 395MS in 10 mM sodium citrate buffer (pH6.0) at 30°C for 10 min. Symbols: ○ 0% BSA; ● 0.2% BSA; ▼ 0.5% BSA; ▲ 1% BSA; ◆ 2% BSA, and ■ 5% BSA.

Effect of temperature. When the bactericidal efficiencies on *S. typhimurium* by decyl (B-10), dodecyl (B-12) and tetradecyl betainate (B-14) were compared at temperatures 0-30°C and pH 4-6 in the presence and absence of a high concentration of BSA (10%), their ranking differed depending on the suspension medium. All compounds were most efficient at 30°C, pH 6 and in the absence of BSA (Fig. 13), B-14 being ca 10 times as efficient as B-12 which, in turn, was ca 10 times as efficient as B-10. These results fit well with earlier data (Fig. 9) indicating that hydrophobic interaction by the hydrocarbon chain is important for killing. For -4 log inactivation ca 5 µg/ml of B-14 was required. At 0°C higher concentrations of each compound were required to achieve the same bactericidal effect, most so for B-14 (50 µg/ml), but the ranking was still found to be the same.

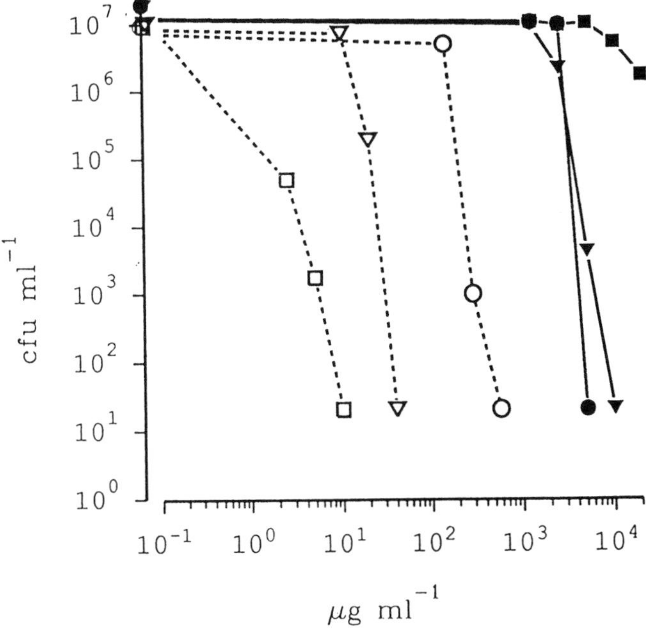

Figure 13. Antimicrobial effects of betaine esters against *S. typhimurium* 395MS in 10 mM sodium citrate buffer (pH 6.0) for 10 min. at 30°C (open symbols), and in the same buffer with 10% bovine serum albumin (BSA) at 0°C (closed symbols). Decyl betainate (B-10) ○ ●, dodecyl betainate (B-12) ▽ ▼, tetradecyl betainate (B-14) ❑ ■.

The reduced antimicrobial effect at lower temperatures might be a combination of decreased hydrophobic interaction and diffusion rate at lower temperatures and the generally observed reduced killing of microbes under conditions of low metabolic activity. Weaker hydrophobic interaction would affect B-14 more than B-12 which might explain the observed differences.

At low temperatures protein extinguishes the activity of BE-14. In the presence of 10% BSA in the suspension medium, 850 μg/ml of B-14 was required at 30°C to cause -4 log inactivation (Fig. 13). On a molar basis, this is 1.9 B-14 molecules per BSA molecule which is close to the 2.1 molecules found earlier. In this experiment, the dilution steps were such that 500 μg B-14 per ml caused no killing effect, whereas at 1000 μg/ml nearly complete killing was achieved. Since in the absence of BSA 10 μg B-14 per ml was enough for complete killing, we conclude that with 500 μg B-14 per ml in the presence of 10% BSA, less than 2% of B-14 is free for antibacterial action. This implies that most of the B-14 is bound to BSA at a 1:1 molar ratio and that at an over-all 2:1 molar ratio (1000 μg B-14 per ml) a minor proportion (1%) is free. The present results may also fit the equation $(A\tau)/(A) = 1 + nKM$ in which $A\tau$ represents total concentration of B-14, A unbound B-14, n binding sites of BSA (M) and K the binding constant[21].

The competitive inhibition of BSA on the bactericidal effect of B-14 was more pronounced than that on B-12 which in turn exceeded that on B-10. At lower temperatures this tendency was enhanced, such that in 10% BSA at 20°C and pH 6 the bactericidal effect of B-10, B-12 and B-14 appeared the same, at 10°C that of B-14 was the weakest, and at 0°C that of B-10 the strongest and that of B-14 inconspicuous (Fig. 13). Weaker competitive inhibition by BSA on B-10 and B-12 is probably a consequence of weaker hydrophobic interaction due to shorter hydrocarbon chains. A similar dependence on chain length has been observed for the inactivation by human serum and yeast cells of the bactericidal effect of C_{12}-, C_{14}-, and C_{16}- benzalkonium chlorides[14]. The exaggeration at lower temperatures of the inhibition of B-14 in comparison to B-12 and B-10 may be a consequence of differences in the relative binding strengths to the bacteria and BSA at different temperatures. Furthermore, in the absence of BSA the bactericidal efficiency of B-14 was reduced more than those of B-12 and B-

10 by reduction of temperature. Moreover, the Krafft point of B-14 occurs at ca 13°C (unpubl.) which leads to a very low free concentration of B-14 at lower temperatures. In the presence of 10% BSA at temperatures below the Krafft point (10°C, 0°C) the free concentration of B-14 is negligible, since the bactericidal effect was nearly inconspicuous. The Krafft points of B-12 and B-10 are below 4°C which seems related to their remaining bactericidal effect at 0°C in the presence of 10% BSA (Fig. 13).

Effect of nonionic amphiphiles

Inhibition of the bactericidal effect of B-14, similar to that caused by BSA, was found, when different concentrations of nonionic detergents were present (Fig. 14). The relationships were similar with the various tested compounds, *viz.* Span 80 (sorbitan monooleate, M ca 460), Triton X-100 (various polyoxyethylene ethers), nonoxynol-10 (polyoxyethylene ethers of nonylphenol, M ca 820), and Tween 80 (oleylsorbitan polyethyleneglycol ether, M ca 1300). In order to confine the killing of *S. typhimurium* to -4 log in the presence of 100 or 1000 µg/ml of B-14, ca 36 times (SD = ± 26%) as much of the nonionic detergents were required on weight basis. On a molar basis this means that 9 (Tween 80) to 25 (Span 80) nonionic amphiphilic molecules neutralized one B-14 molecule. Most likely, these results are a consequence of the incorporation of B-14 into mixed micelle arrangements with the nonionic amphiphiles. This is supported by the finding that the amount of free quaternary ammonium compound in the presence of polysorbate 80 (Tween 80) is very small until a critical concentration has been reached[22] which may account for the absence of antimicrobial effect.

In the presence of 0.2% BSA in different buffers, *S. typhimurium*, *P. aeruginosa* and *C. jejuni* were inactivated more extensively by B-14 than by B-16 and at pH 6 than at pH 7 (Table 1). These results extend the earlier observations (Figs. 11 & 13) to more microbes and buffers and support the universal character of the phenomenon. These and other (Table 2) results are also consistent with the observations that Ca^{++} stabilizes the outer membrane of *P. aeruginosa*, and that the bacteria become more sensitive to quaternary ammonium componds after removal of Ca^{++} [23].

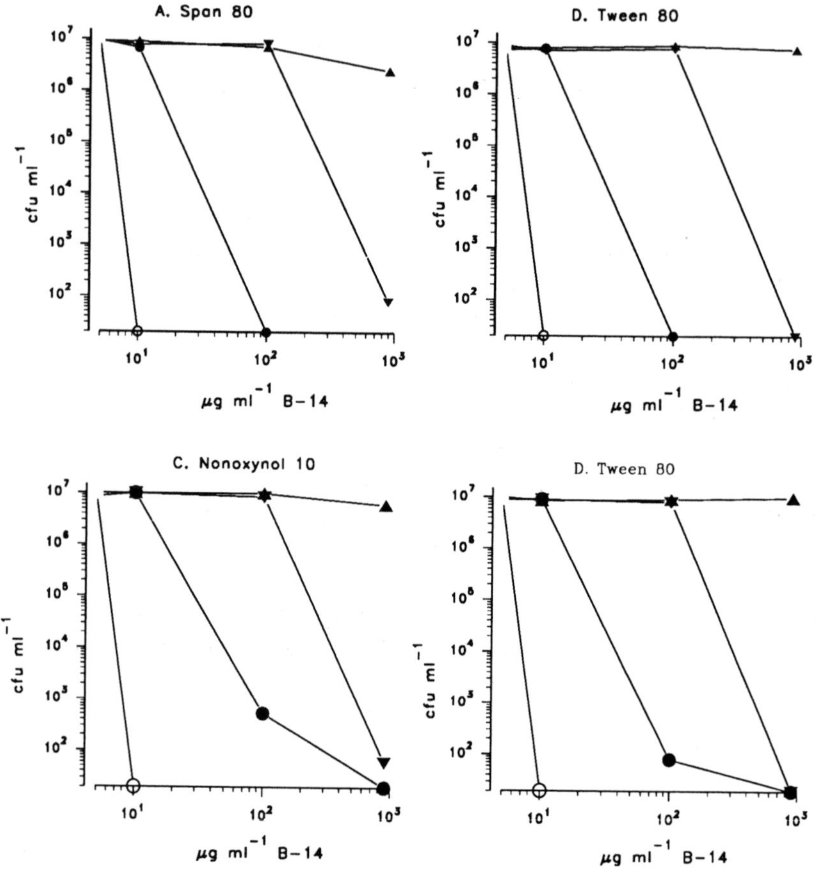

Figure 14. Effect of nonionic detergents on the bactericidal effect of tetradecyl betainate (B-14) against *S. typhimurium* 395MS in 10 mM sodium citrate buffer (pH 6.0) at 30°C for 10 min. A. Span 80 (sorbitan monooleate); B. Triton X-100 (polyoxyethylene ethers); C. Nonoxynol 10 (polyoxyethylene ethers of nonylphenol); D. Tween 80 (oleylsorbitan polyethyleneglycol ether). Symbols: ○ 0%, ● 0.2%, ▼ 2%, ▲ 20% of non-ionic detergents.

Table 2. Inactivation of *P. aeruginosa* by 10 or 30 µg/ml tetradecyl (B-14) or hexadecyl betainate (B-16) in 10 mM sodium citrate buffer pH 6 and in 100 mM tris buffer pH 7 at 30°C for 10 min.

	citrate pH 6		tris pH 7	
	B-14	B-16	B-14	B-16
betainate 10 µg/ml	>-3.7	>-3.7	-0.8	-1.7
betainate 30 µg/ml	>-3.7	>-3.7	-3.1	-1.4

Effects on the Envelope of *E.coli*.

There was a selective release of the periplasmic enzyme β-lactamase from *E.coli* (5 x 10^9 cfu/ml) as compared to cytoplasmic β-galactosidase at a narrow concentration interval of B-14 (250-1000 µg/ml). At low concentrations of B-14 (250 µg/ml), higher enzyme activities were released at alkaline (pH 9) than at acid (pH 5) pH values (Table 3).

Table 3. Release of β-lactamase activity, after exposure of *E. coli* to tetradecyl betainate (B-14) at pH 5, 7 and 9.

Proto-col[a]	pH	β-lactamase activity* B-14 250 µg/ml	B-14 1000 µg/ml
A	5	0	2.08
	7	3.52	0.96
	9	5.28	0.48
B	5	0.32	18.72
	7	1.6	14.08
	9	2.56	13.76

* Absorbance at 482 nm after hydrolysis of Nitrocefin.
[a]A = suspension medium maintained throughout experiment. Neutralization before determination of enzyme activity.
B = one min. after exposure to B-14 adjusted to 2M NaCl, pH 9. Kept for 30 min. until neutralization.

This may be due to enhanced ionization of the bacterial surface at pH 9 with facilitated charge interaction and similar to the greater bactericidal effect at higher pH (*vide supra*). At higher concentrations of B-14 (1000 μg/ml) and pH 7-9, the yield of β-lactamase was smaller than at 250 μg/ml probably as a consequence of enzyme denaturation by B-14. At this concentration, higher enzyme activities were obtained at pH 5 than at pH 9 indicating that enzyme denaturation is promoted at higher pH by stronger charge interaction similar to the greater inhibition by BSA of the bactericidal effect of B-14 at higher pH. Considerably higher yields were achieved, when, after a short exposure to 1000 μg/ml, the suspension was adjusted to 2M NaCl and pH 9 in order to reduce charge interaction and promote hydrolysis of B-14.

Effects on the mammalian mucosa.

Quaternary ammonium compounds were injected into rat small intestinal ligated loops. Some 5 min. after the injection, there was an extensive shedding of the enterocytes from the villi, leaving the basement membrane denuded (Fig. 15 A-E). Thirty min. after challenge, most of the enterocytes could be found in the intestinal lumen, goblet cells covering the now flattened villi. Concomitant with the shedding of the enterocytes, clear liquid was secreted into the intestinal lumen and the permeability was affected. It must be stressed that there were no bleeding or inflammatory reaction in the submucosal tissues. A smaller effect was seen with B-14 than with the stable quaternary ammonium compound cetyltrimethylammonium bromide (Table 4) indicating hydrolysis of B-14. Both compounds left the basement membrane intact. In contrast, the nonionic amphiphile nonoxynol-10, in addition to the above effects, caused non-selective breakdown of the intestinal wall including the basement membrane. This implies a potential risk for application of nonoxynol on mucosal membranes containing this type of epithelium.

Table 4. Secretion in rat intestinal ligated loop (mg weight increase/cm length of intestine) 6 h after injection of 2,5 and 10 mg amphiphilic compound in one ml PBS.

	Net fluid accumulation (mg/cm) mg per loop injected		
compound	2	5	10
tetradecyl betainate (B-14)	12±2	25±11	199±8
cetyltrimethylammonium bromide (CTAB)	27±9	171±4	456±12
nonoxynol-10	25±7	98±6	422±4

Surface disinfection of chicken.

Poultry is often contaminated with *Salmonella* spp. and *Campylobacter jejuni* which are frequent sources of enteritis in humans. Since these bacteria are heat-sensitive, and poultry is cooked before consumption, the routes of transmission are considered to mainly occur by surface contact in the kitchen with contaminated food which is consumed without cooking. Both B-12 and B-14 caused rapid surface disinfection of chicken contaminated with *C. jejuni* and *S. typhimurium,* more effectively at 37°C than at 14°C. At 37°C, 0.1% B-12 caused more than 99% killing in 10 min. of *C. jejuni* as well as *S. typhimurium,* whereas at 14°C 1% B-12 was required for 99% inactivation of *C.jejuni* and 2% for *S.typhimurium*. Since these substances spontaneously hydrolyze into betaine and fatty alcohols, more rapidly at higher temperatures, this procedure might be exploited to reduce the risk for transmission of pathogenic bacteria in the handling of poultry.

Inactivation of sexually transmitted agents and other genital microbes.

Spermicidal creams, gels, foams etc are often used as chemical aids to mechanical barrier contraceptives. Many spermicidal preparations also inactivate sexually transmitted disease (STD) agents *in vitro,* so they might

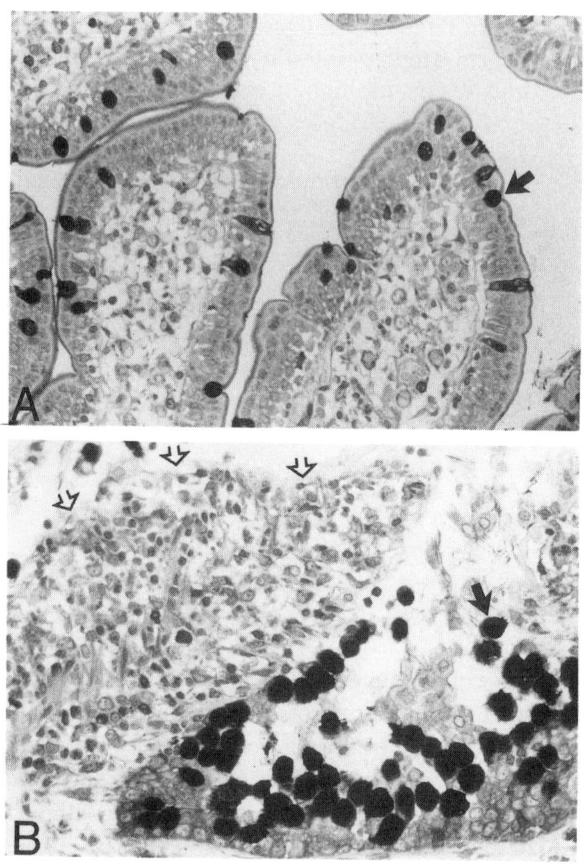

Figure 15. Plastic sections from rat small intestine stained with PAS to demonstrate goblet cells, marked by black arrow in all micrographs. A. Normal animal. Enterocytes dominate the mucosal surface. Only few goblet cells. B. Intestinal loop exposed to 5 mg of B-14 for 5 min. Note extensive shedding of enterocytes. Open arrows indicate denuded basement membrane. C. Intestinal loop exposed to B-14 for 15 min. Open arrow indicates detached cells in the intestinal lumen. D. Intestinal loop exposed to B-14 for 30 min. Enterocytes have been shed. The basement membrane is covered by goblet cells. E. Intestinal loop exposed to B-14 for 60 min. Goblet cells cover most of the villi. Note absence of inflammatory reaction.

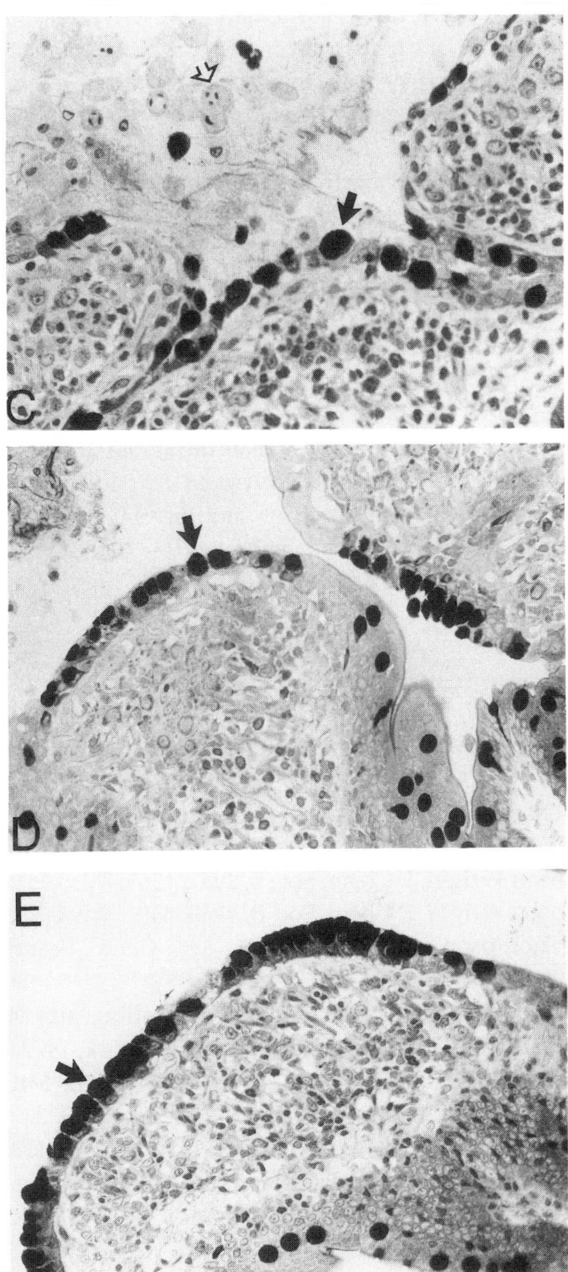
C
D
E

help to prevent the spread of STD. The non-ionic detergent nonoxynol is the active ingredient in most spermicides. Since this substance is lytic for mammalian tissue and rather resistant against biological degradation, actions have been initiated to minimize its use.

The *in vitro* microbicidal activity of B-14 against *Chlamydia trachomatis, Gardnerella vaginalis, Haemophilus ducreyi, Neisseria gonorrhoeae, Streptococcus agalactiae, Candida albicans, Herpes simplex virus (HSV),* and *Human immunodeficiency virus (HIV)* was fast and similar to that of stable quaternary ammonium compounds. As compared to nonoxynol, quaternary ammonium compounds are active against a wider range of microbes. Furthermore, quaternary ammonium compounds seem to be less destructive to mucosal membranes than nonoxynol. Since the betaine esters seem to irritate the mucosal membranes less than the stable quaternary ammonium compounds and probably hydrolyse *in situ* to ordinary human metabolites they may be considered as antimicrobials for use in close association with the genital mucous membranes.

REFERENCES

1. N. Bodor, Med. Res. Rev., 1984, 4, 449.
2. M. Lindstedt, S. Allenmark, R. A. Thompson and L. Edebo, Antimicrob. Agents Chemother, 1990 34, 1949.
3. R. A. Thompson, M. Lindstedt and S. Allenmark, Anal. Lett.,1990, 23, 787 .
4. M. Robson Wright, J. Chem. Soc. B, 1967, 1265.
5. M. Robson Wright, J. Chem. Soc. B, 1968, 548.
6. R. A. Thompson and S. Allenmark, Acta Chem. Scand, 1989, 43, 690.
7. L. S. Romsted, in 'Micellization, Solubilization, and Microemulsions' (K. L. Mittal, Ed.). Plenum Press, New York, 1977. Vol. 2.
8. L. S. Romsted, in 'Surfactants in Solution' (K. L. Mittal and B. Lindman, Eds.). Plenum Press, New York, 1984, Vol. 2.
9. C. A. Bunton and G. Savelli, Adv. Phys. Org. Chem., 1986, 22, 213.
10. C.A. Bunton, 'Kinetics and Catalysis in Microheterogeneous Systems', (M. Grätzel and K Kalyanasundaram Eds.), Marcel Dekker, New York, 1991, Chapter 2, p. 13.

11. R.A. Thompson and S. Allenmark, J. Colloid Interface Sci., 1992, in press.
12. C. A. Bunton and B. Wolfe, J. Amer. Chem. Soc., 1973, 95, 3742.
13. D. Bartet, C. Gamboa and L. Sepúlveda, J. Phys. Chem., 1980, 84, 272.
14. J.J. Merianos, 'Quaternary ammonium antimicrobial compounds'. In S. Block, 'Disinfection, Sterilization and Preservation'. Lea & Febiger, Philadelphia & London, 1991, pp. 225-255. ISBN 0-8121-1364-0.
15. W.A. Hamilton, FEBS Symp., 1970, 20, 71.
16. R.E.W. Hancock, Annu. Rev. Microbiol., 1984, 38, 237.
17. W.B. Hugo, 'Disinfection mechanisms'. In A.D. Russel, W.B. Hugo, and G.A.J. Ayliffe (ed.), Principles and practice of disinfection, preservation and sterilisation. Blackwell Scientific Publications, Ltd., Oxford, 1982, pp 158-185.
18. J.E. Cronan, and C.O. Rock, 'Biosynthesis of membrane lipids', In F.C. Neidhardt, J.L. Ingraham, K.B. Low, B. Magasanik, M. Schaechter, and H.E. Umbarger (ed.), vol. 1. American Society for Microbiology, Washington, D.C., 1987, pp 474-497.
19. H.W. Wollenweber, K.W. Broady, O. Lüderitz, and E.T. Rietschel, Eur. J. Biochem., 1982, 124, 191.
20. S.P. Denyer, Mechanism of action of biocides, Internat. Biodeter., 1990, 26, 89.
21. H.B. Kostenbauder, 'Physical factors influencing the activity of antimicrobial agents'. In S. Block, 'Disinfection, Sterilization, and Preservation'. Lea & Febiger, Philadelphia & London, 1991, pp. 59-71.
22. P.P. DeLuca, and H.B. Kostenbauder, J. Am. Pharm. Assoc., Sci. Ed., 1960, 49, 430.
23. M.R.W. Brown, and R.M.E. Richards, Nature, 1965, 207, 1391.

Production and Application of Sulphobetaines

W. Kurze

RASCHIG AG, 6700 LUDWIGSHAFEN, GERMANY

1 PREFACE

Raschig AG which I speak for here today are newcomers in the field of tensides. We are not in the market for the production of detergents but rather offer special systems for use with specific tensides. The products of this chemistry are mainly used in the field of electrochemical additives and emulsion polymerization.

2 DEFINITION

Sulphobetaines represent compounds containing one sulphonate anion and one heteroatom cation combined in one single molecule.

General formula of sulphobetaine

$$R_2-\overset{\displaystyle R_1}{\underset{\displaystyle R_3}{X^{\oplus}}}-R_4-SO_3^{\ominus}$$

X = N will yield ammonium sulphobetaine and X = P phosphonium sulphobetaine.
Since all of the Rs may be varied, there's a large number of compounds feasible, which partly have been synthesized already.

3 INTRODUCTION

The development of new substances is essentially influenced by three factors:
a) interest of the market in the task

b) research interest

c) economic production

For many years, industry has been interested in zwitterionic compounds since they combine cationic and anionic functions in one single molecule. According to our knowledge, the substances today's industry is particularly interested in may be described by the following formula:

$$R_2-\overset{\overset{R_1}{|}\oplus}{\underset{\underset{R_3}{|}}{N}}-CH_2-COO^{\ominus} \quad (1) \qquad \text{and} \qquad R_2-\overset{\overset{R_1}{|}\oplus}{\underset{\underset{R_3}{|}}{N}}-(CH_2)_m-O_n-SO_3^{\ominus} \quad (2)$$

Figure 1 Economically significant zwitterionic compounds

In this formula, R_1 and R_3 may be methyl, R_2 may be a fatty alkyl or a fatty alkoylamidoalkyl, with m most likely being 2 or 3 and n = 0 or 1. The R_2 variants particularly provide for the lipophilic character to be varied, while n controls the distance of the ionic centers, and m the resistance to hydrolysis. m = 0 will yield sulphonic acids; they are stable as far as hydrolysis is concerned, while m = 1 will yield sulphates, i.e. esters which are not resistant to hydrolysis.

In water and following hydrolysis, they will loose their anionic center according to the formula below:

$$R_2-\overset{\overset{R_1}{|}\oplus}{\underset{\underset{R_3}{|}}{N}}-(CH_2)_m-O-SO_3^{\ominus} \ (3) + H_2O \longrightarrow R_2-\overset{\overset{R_1}{|}\oplus}{\underset{\underset{R_3}{|}}{N}}-(CH_2)_m-OH + HSO_4^{\ominus} \ (4)$$

Figure 2 Hydrolysis of betaine sulphates

In this paper, we will not discuss substance class (1), since the weaker anion center in particular will frequently fail to perform as well as does the sulphonate, although it may be produced easily.

Substance class (2) is most frequently produced according to the following principles:

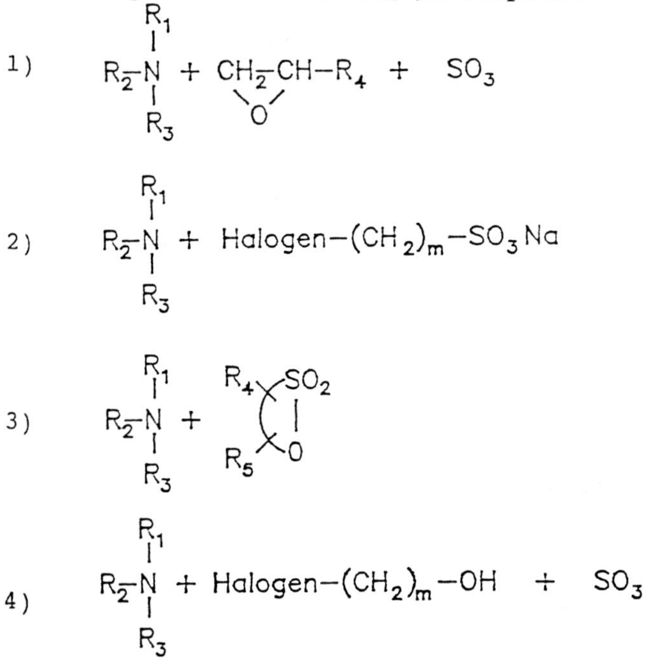

1) $$R_2{-}N(R_1)(R_3) + CH_2{-}CH{-}R_4 \text{ (epoxide, } {-}O{-}) + SO_3$$

2) $$R_2{-}N(R_1)(R_3) + \text{Halogen}{-}(CH_2)_m{-}SO_3Na$$

3) $$R_2{-}N(R_1)(R_3) + \text{sultone } (R_4, R_5; {-}SO_2{-}O{-})$$

4) $$R_2{-}N(R_1)(R_3) + \text{Halogen}{-}(CH_2)_m{-}OH + SO_3$$

Figure 3 Principles most frequently used for the production of sulphobetaines

Principle 1) requires the use of a special technology (ethylene oxide, propylene oxide, sulphur trioxide).

Principle 2) will cause chlorides to be produced as a byproduct.

Principle 4) will also cause sodium chloride to be generated as a byproduct.

Only principle 3) using alkane sultones, e.g. propane-1,3-sultone will provide for the desired betaine to be produced without generating sodium chloride or other byproducts.

4 SULPHOPROPYLATIONS

The sultones represent alkylation agents enabling the introduction of the sulphoalkyl groups. In the presence of acid H-atoms together with alkali, the alkylation reaction will take place as a nucleophilic substitution, or in the case of tertiary amines or tertiary phosphines will take place as mere addition. In the first case, the alkali salts of the desired compounds are generated, from which the dangling acids may be produced if necessary.

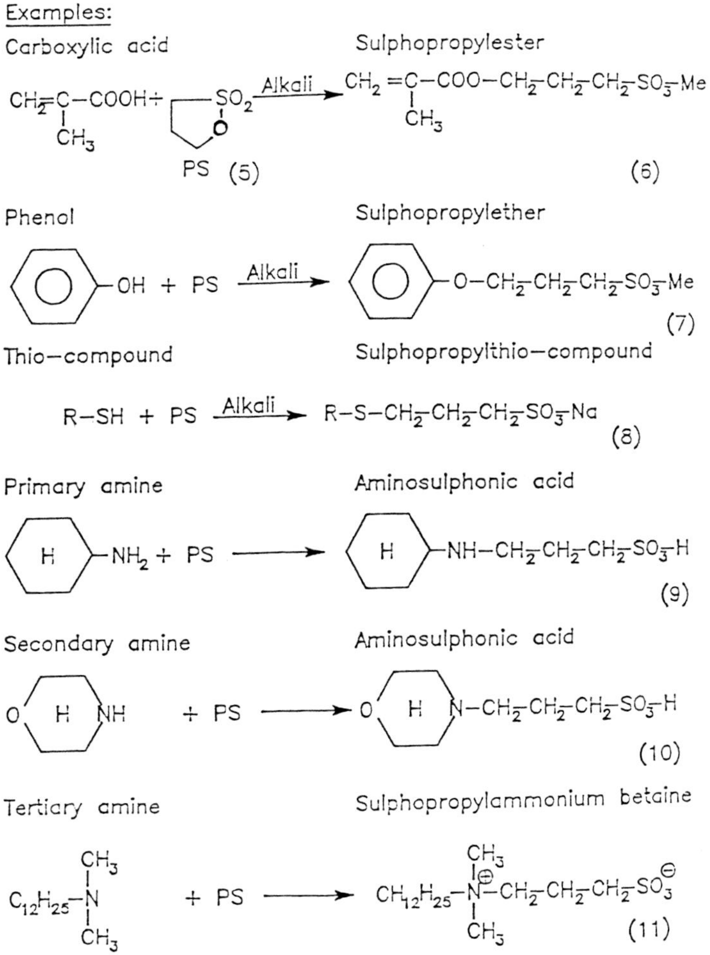

Examples:

Carboxylic acid → Sulphopropylester

$$CH_2{=}C(CH_3){-}COOH + \text{PS (5)} \xrightarrow{\text{Alkali}} CH_2{=}C(CH_3){-}COO{-}CH_2{-}CH_2{-}CH_2{-}SO_3{-}Me \quad (6)$$

(PS = propane sultone, cyclic $-SO_2-O-$)

Phenol → Sulphopropylether

$$C_6H_5{-}OH + PS \xrightarrow{\text{Alkali}} C_6H_5{-}O{-}CH_2{-}CH_2{-}CH_2{-}SO_3{-}Me \quad (7)$$

Thio-compound → Sulphopropylthio-compound

$$R{-}SH + PS \xrightarrow{\text{Alkali}} R{-}S{-}CH_2{-}CH_2{-}CH_2{-}SO_3{-}Na \quad (8)$$

Primary amine → Aminosulphonic acid

$$C_6H_{11}{-}NH_2 + PS \longrightarrow C_6H_{11}{-}NH{-}CH_2{-}CH_2{-}CH_2{-}SO_3{-}H \quad (9)$$

Secondary amine → Aminosulphonic acid

$$O(C_4H_8)NH + PS \longrightarrow O(C_4H_8)N{-}CH_2{-}CH_2{-}CH_2{-}SO_3{-}H \quad (10)$$

Tertiary amine → Sulphopropylammonium betaine

$$C_{12}H_{25}{-}N(CH_3)_2 + PS \longrightarrow C_{12}H_{25}{-}N^{\oplus}(CH_3)_2{-}CH_2{-}CH_2{-}CH_2{-}SO_3^{\ominus} \quad (11)$$

Figure 4: Sulphopropylation reaction

Me = metal

Functionally speaking, the introduction of the sulphopropyl chain will

1) increase the solubility in water
2) enlarge the molecule
3) introduce an anionic hydrophilic remainder.

5 SULPHOBETAINES

The term 'sulphobetaines' in this paper relates to the real sulphobetaines described by the formula

$$\oplus\!-\!SO_3^{\ominus} \quad (12)$$

rather than the sulphobetaines simply produced in the course of a tautomerism, such as aminosulphonic acids.

$$NH_2-R-SO_3H \rightleftharpoons \overset{\oplus}{N}H_3-R-SO_3^{\ominus} \quad (13)$$

We have concentrated our efforts on the production of ammonium sulphobetaines:

$$\begin{matrix} R_1 \\ R_2 \\ R_3 \end{matrix} \Big> \overset{\oplus}{N}-R_4-SO_3^{\ominus} \quad (14)$$

R_1, R_2 and R_3 may be alkylic groups of C_1-C_{20}.

R_1, R_2, R_3, N may be alkanoylamido alkyl.
N-hetero cycles, e.g. pyridine,
quinolines,
imidazoles,
other N-hetero cycles.

R_4 may be a straight or branched alkylene chain of C_2-C_{20}.

Let's take the example of pyridine as an amine to list some partly new sulphoalkylic compounds:

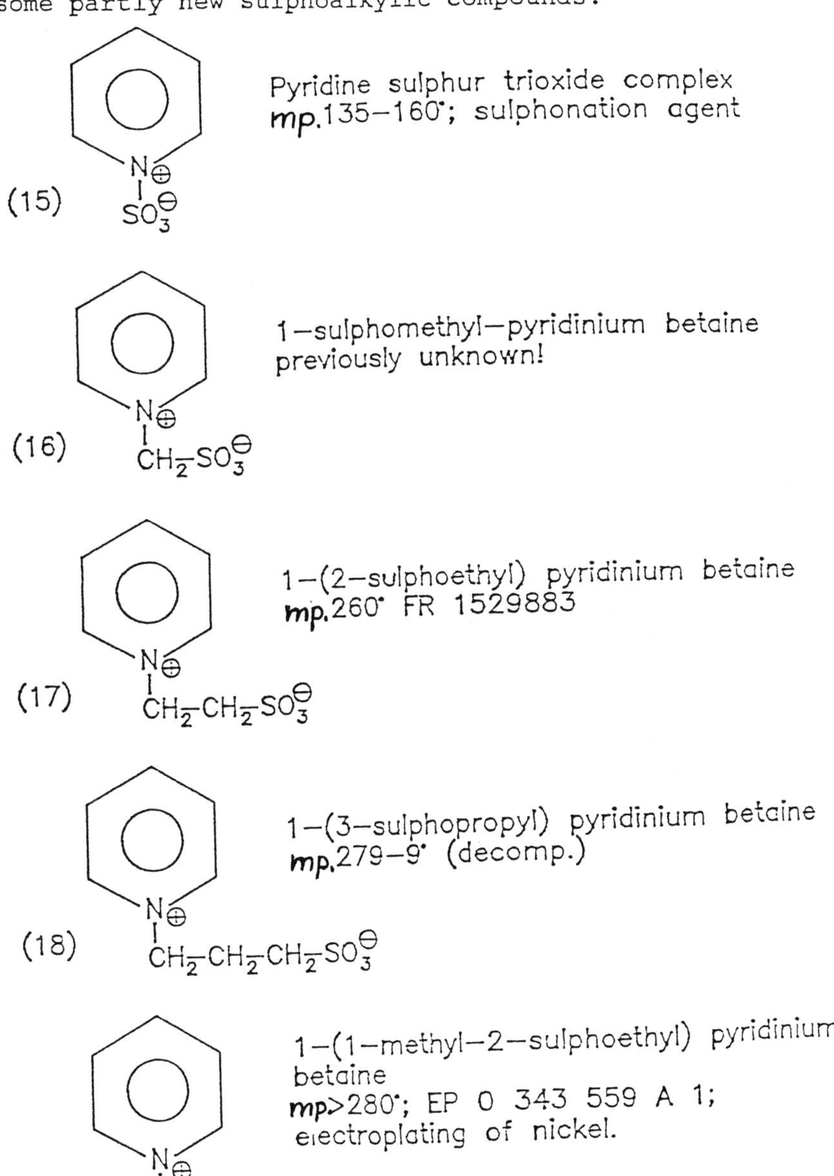

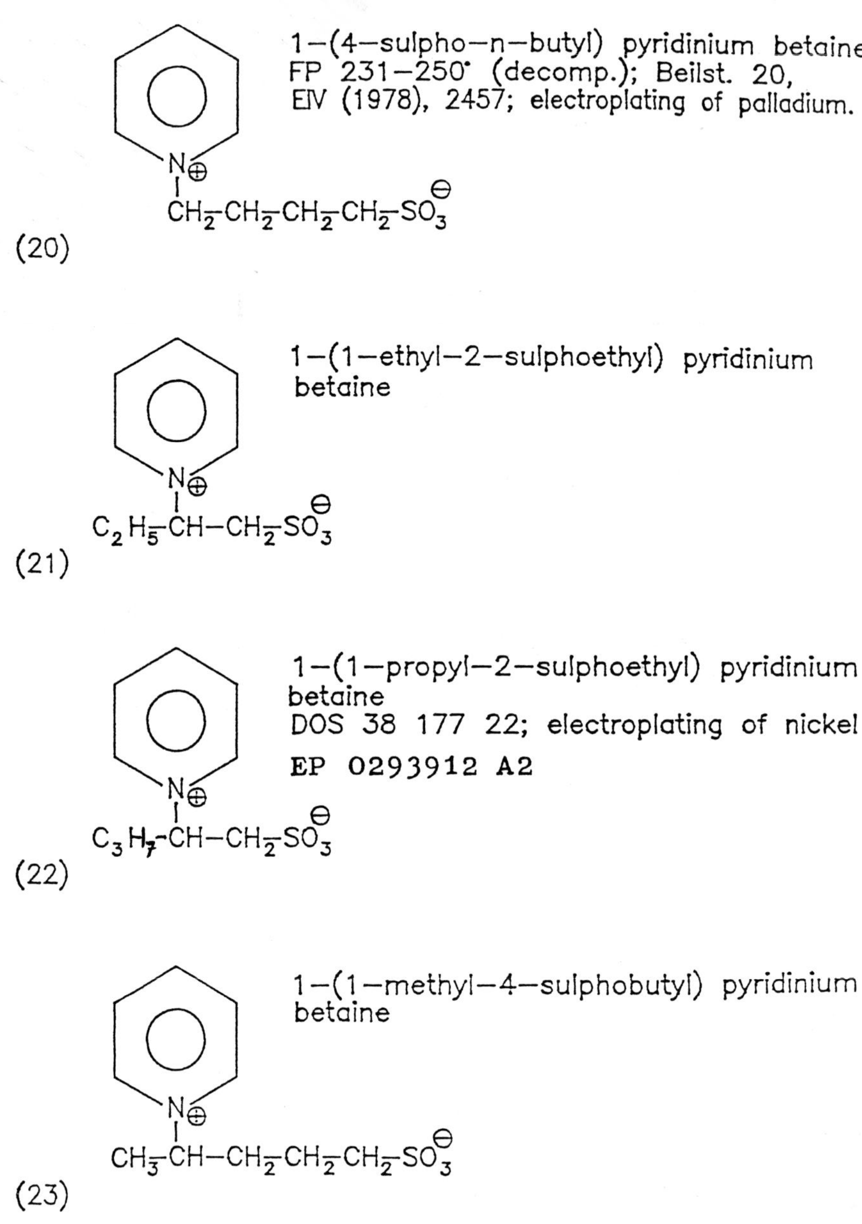

Figure 5: 1–sulphoalkyl pyridinium betaines

Selecting the relevant sultone

C_2	carbylsulphate	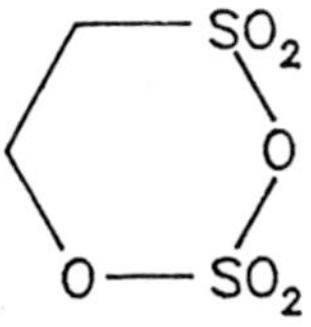	(24)
C_3	propane-1,2–sultone unstable	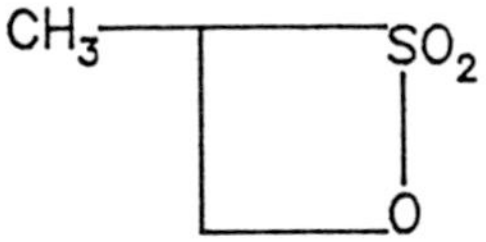	(25)
	propane–1,3–sultone	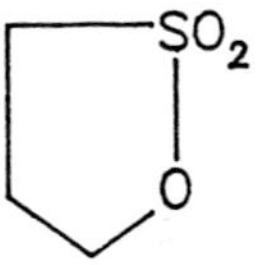	(5)
C_4	butane–1,2–sultone unstable	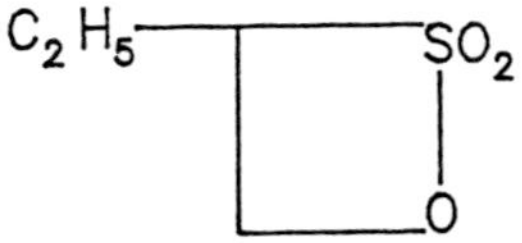	(26)
	butane–1,4–sultone	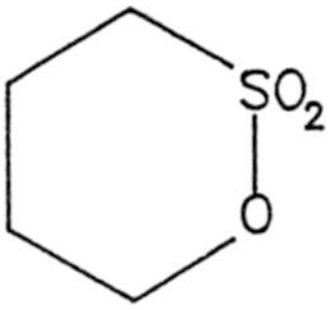	(27)
C_5	pentane–1,2–sultone unstable	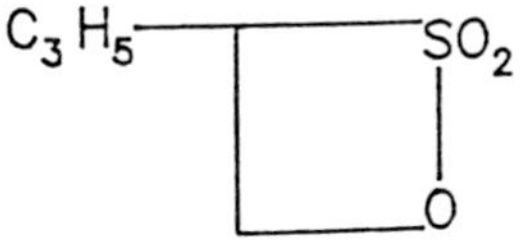	(28)
	pentane–1,4–sultone	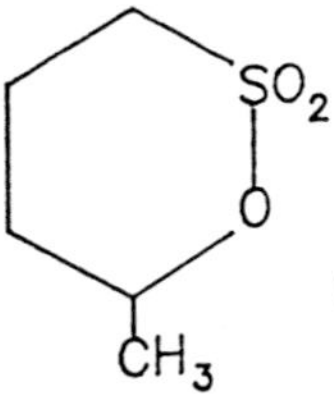	(29)

Figure 6: Several sultones

provides for the sulphonate half of the molecule to be varied strongly.

6 SULPHOPROPYLBETAINE

For many years, Raschig AG have been engaged in the production of sulphopropyl derivatives. Here, the large number of tertiary amines and N-containing hetero cycles available provide access to a large range of sulphobetaines.
In the following, some of these compounds and their typical applications will be listed.

SPE and SPV

SPE

$$CH_2{=}C(CH_3){-}COO{-}CH_2{-}CH_2{-}N^{\oplus}(CH_3)_2{-}CH_2{-}CH_2{-}CH_2{-}SO_3^{\ominus} \quad (30)$$

SPV

$$\text{2-}(CH{=}CH_2)\text{-pyridinium-}N^{\oplus}{-}CH_2{-}CH_2{-}CH_2{-}SO_3^{\ominus} \quad (31)$$

The products SPE and SPV represent functional monomers. In aqueous emulsion polymerization, they partly or wholly assume the task of micelle formation. As soon as the process of polymerization has been completed, the monomer will be firmly integrated in the molecular structure. Consequently, the films of such systems stand out for their high initial water resistance.

Other substances featuring a similar chemistry, such as SPA and SPM also represent new functional monomer sulphopropyl compounds which, however, are no betaines.

$$CH_2{=}C(R){-}COO{-}CH_2{-}CH_2{-}CH_2{-}SO_3K \quad (32)$$

R	compound
H	SPA
CH_3	SPM

3-formyl-1-(3-sulphopropyl)pyridinium betaine

Also, an aldehyde group in the compound (33) may serve as redox partner or function to continue the expansion using amines.

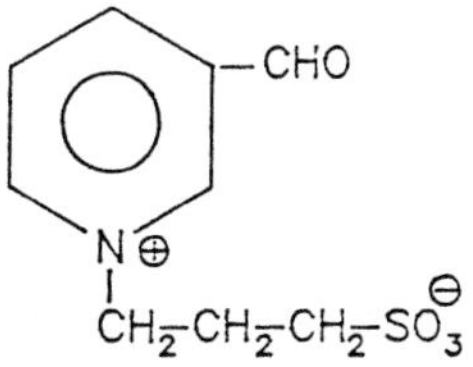

$$\text{3-CHO-C}_5\text{H}_4\text{N}^{\oplus}\text{-CH}_2\text{-CH}_2\text{-CH}_2\text{-SO}_3^{\ominus} \quad (33)$$

PPS and PPSOH

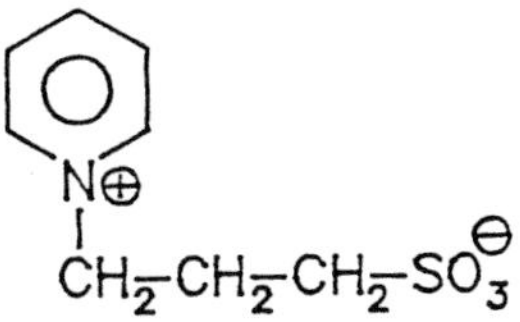

$$\text{C}_5\text{H}_5\text{N}^{\oplus}\text{-CH}_2\text{-CH}_2\text{-CH}_2\text{-SO}_3^{\ominus} \quad (18)$$

$$\text{C}_5\text{H}_5\text{N}^{\oplus}\text{-CH}_2\text{-CH(OH)-CH}_2\text{-SO}_3^{\ominus} \quad (34)$$

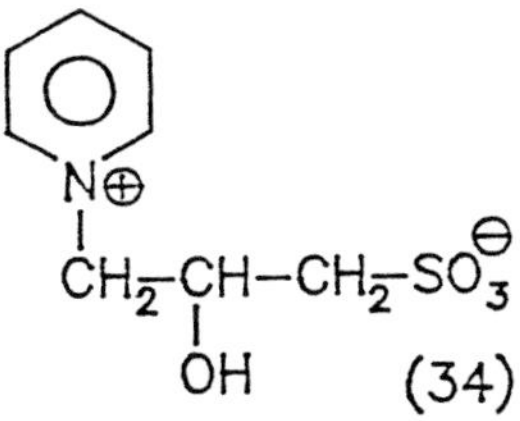

Both substances are successfully used as brighteners for galvanic nickel baths. The function of brighteners is very complex. Apart from a small number of tenside functions, there are other processes and secondary reactions involved. It is even presumed that they are incorporated in the metal layer.

Ralufon[R] betaines

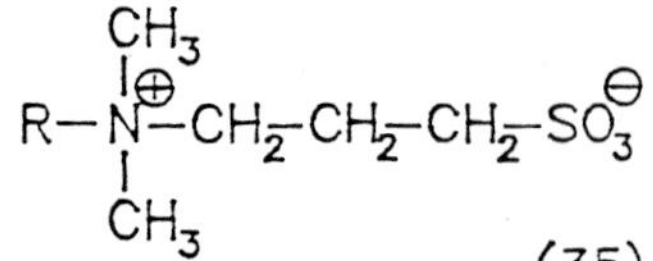

$$R-\overset{\overset{CH_3}{|}}{\underset{\underset{CH_3}{|}}{N^{\oplus}}}-CH_2-CH_2-CH_2-SO_3^{\ominus} \quad (35)$$

R = $C_{12}H_{25}$ Ralufon DCH, DL
= $C_{14}H_{29}$ Ralufon DM
= $C_{16}H_{33}$ Ralufon DP
= $C_{18}H_{37}$ Ralufon DS
= $C_{18}H_{35}$ Ralufon DT
= $C_{11}H_{23}$-CO-NH-$(CH_2)_3$ Ralufon CA
= $C_{17}H_{35}$-CO-NH-$(CH_2)_3$ Ralufon TA

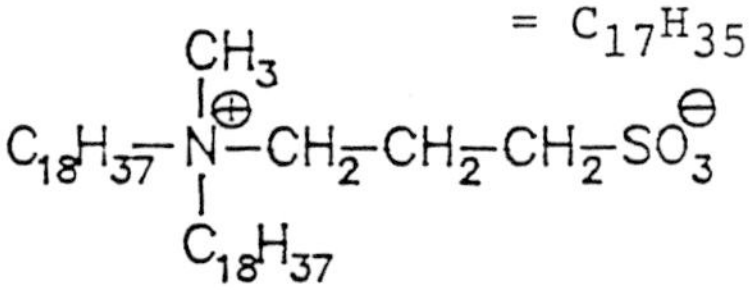

$$C_{18}H_{37}-\overset{\overset{CH_3}{|}}{\underset{\underset{C_{18}H_{37}}{|}}{N^{\oplus}}}-CH_2-CH_2-CH_2-SO_3^{\ominus} \quad (36)$$

Ralufon MDS

Figure 7 Ralufon[R] betaines

Of the hydroxylated type (38) too,

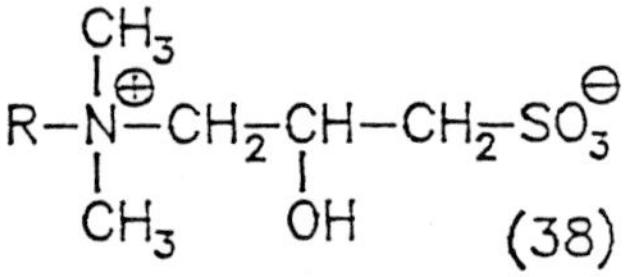

$$R-\overset{\overset{CH_3}{|}}{\underset{\underset{CH_3}{|}}{N^{\oplus}}}-CH_2-\underset{\underset{OH}{|}}{CH}-CH_2-SO_3^{\ominus} \quad (38)$$

$$R-CO-NH-(CH_2)_3-\overset{\overset{CH_3}{|}}{\underset{\underset{CH_3}{|}}{N^{\oplus}}}-CH_2-\underset{\underset{OH}{|}}{CH}-CH_2-SO_3^{\ominus} \quad (39)$$

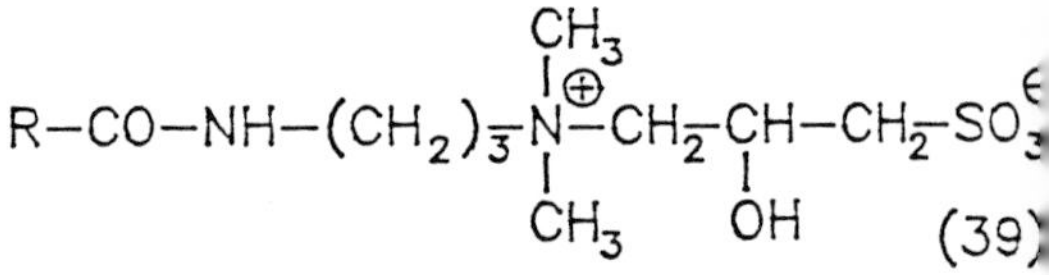

there is a large number of preparations available on the market.

The type to be mentioned in particular is the cocoamidopropyldimethylammonium derivative (39), which is used in shampoos, disinfectants and body cleaners. The Ralufon-betaines are extremely efficient tensides. They may be used within a wide pH-range - from mineral acid up to neutral media, and partly in alkaline solutions such as hypochlorite solution. Moreover, there were no discolorations observed when used with bleaching agents.

When made from sultones, they are free from mineral salts. In the first place, they are resistant to hydrolysis, which makes them particularly suited for applications involving long dwell times (technical cleaners, plating industry) or high temperatures (tertiary crude oil production). They are also operative in hard water, or partly in saline media (tertiary crude oil production). Some of the substances are expected to be bactericidal.

Also, the tensides are compatible with other cationic or anionic tensides.

Applications range from technical special tensides to lime soap dispergators [1], shampoos [2], emulsifying agents, emulsion polymer, neutral cleaners and household or sanitary cleaners.

These tensides are perfectly suited for special fields of application, since this class of substances is highly variable and therefore is easily adapted to the requirements of the user.

<u>Benefits offered for the application of sulphobetaine tensides</u>

- Wide pH-range
- Compatibility with other anionic, cationic and non-ionic tensides
- Resistant to hydrolysis
- Temperature resistant
- Resistant to lime
- Partly free from mineral salts
- Partly incorporated in the polymerization process - favourable in respect of pollution load
- Partly operative in saline media
- Partly AOX-free

<u>Table 1</u>

Surface tension in mN/m of several betaines observed in aqueous solutions of 0.1% at 25°C.

De-ionized water = DW		66
Town water = TW		60
3-formyl-1-(3-sulphopropyl)pyridinium betaine	(DW)	63
N,N-dimethyl-N-methacrylamidopropyl-N-(3-sulphopropyl)ammonium betaine	(DW)	61
1-(2-hydroxy-3-sulphopropyl)pyridinium betaine	(DW)	59
2-methyl-1-(3-sulphopropyl)pyridinium betaine	(DW)	59
3-methyl-1-(3-sulphopropyl)pyridinium betaine	(DW)	59
N,N-dimethyl-N-methacryloxyethyl-N-(3-sulphopropyl)ammonium betaine	(DW)	59
3-carbomoyl-1-(3-sulphopropyl)pyridinium betaine	(DW)	58
4-methyl-1-(3-sulphopropyl)pyridinium betaine	(DW)	58
N,N-dimethyl-N-stearyl-N-(3-sulphopropyl)ammonium betaine	(DW)	38
N,N-dimethyl-N-tallow fatty acid amidopropyl)-N-(3-sulphopropyl)ammonium betaine	(DW)	37
N,N-dimethyl-N-palmityl-N-(3-sulphopropyl)-ammonium betaine	(DW)	37
N,N-dimethyl-N-lauryl-N-(3-sulphopropyl)-ammonium betaine	(DW)	37
N,N-distearyl-N-methyl-N-(3-sulphopropyl)-ammonium betaine	(DW)	36
N,N-coconut fatty acid amidopropyl-N,N-dimethyl-N-(3-sulphopropyl)-ammonium betaine	(DW)	35
N,N-coconut oil-N,N-dimethyl-N-(3-sulphopropyl)-ammonium betaine	(DW)	34
Lauryl sulphate	(TW)	31
Fluorad^{R} FC 129 of 3 M	(TW)	18

Table 2 Table of surface tensions

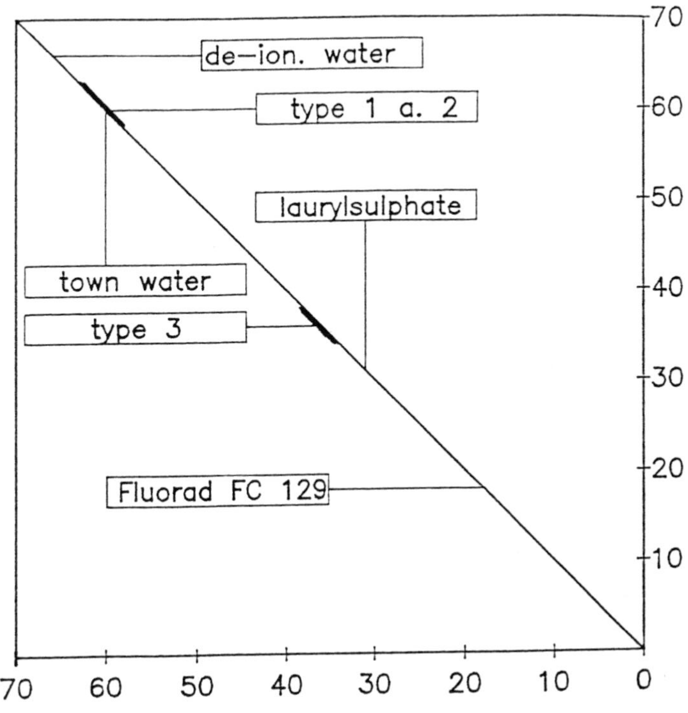

Concentration : 0,1% in water
Appliance : NIMA TS 900
Method : Wilhelmy principle

Figure 8a Surface tensions of several sulphopropyl–ammonium betaines in mN/m

Type 1

$$\text{Pyridinium}(R_1)-N^{\oplus}-CH_2-CH(R_2)-CH_2-SO_3^{\ominus}$$

Compound	R_1	R_2
(18)	H	H
(34)	H	OH
	2-CH_3	H
	3-CH_3	H
	4-CH_3	H
(31)	3-CHO	H
	3-$CONH_2$	H
(30)	2-$CH{=}CH_2$	H

Type 2

$$CH_2{=}C(CH_3)-CO-R-(CH_2)_2-N^{\oplus}(CH_3)_2-(CH_2)_3-SO_3^{\ominus}$$

R
-O-
-NH-CH_2-

Type 3

$$R_2-N^{\oplus}(R_1)(CH_3)-(CH_2)_3-SO_3^{\ominus}$$

Compound	R_1	R_2
(35)	CH_3	$C_{12}H_{25}$-
(35)	CH_3	$C_{16}H_{33}$-
(35)	CH_3	$C_{18}H_{35}$-
(35)	CH_3	$C_{11}H_{23}$-CONH-$(CH_2)_3$-
(35)	CH_3	$C_{17}H_{35}$-CONH-$(CH_2)_3$-
(35)	$CH_{18}H_{37}$	$C_{18}H_{37}$

SLS = sodium lauryl-sulphate
FC = potassium perfluoro-alkylcarboxylate

(Fluorad[R] FC 129 of 3M)

Figure 8b Surface tensions of several sulphopropyl-ammonium betaines.

As far as pollution load is concerned, two aspects should be mentioned. In many places, chlorine chemistry has become problematic due to AOX residues, i.e. remainders of organically bound chlorine in the sewage. The PS betaines referred to above are practically free from chlorine and will not allow for additional AOX loading to occur. In contrast to this, the betaines obtained from halogen alkane sulphonates will frequently contain AOX residues.

Another aspect that should be considered is the biological degradability of the substances in water. The betaines featuring aliphatic alkyl chains and ammonium, sulphonate, sulphate and carbonamide groups are degradable. They do not contain any isononyl phenols which are known to be less degradable.

A degradation of 70 - 90 % of primary C_{10}-C_{16}-alkyl sulphonates was reported by Köbel[3] and Cardini[4], and a DOC degradation of 95 - 100 % reported by Cordon[5] for the same class of substances. In addition, a degradation of 70 - 100 % was reported for α-olefin sulphonates, depending on the method applied[6]. A brief summary of degradation rates of alkyl sulphonates and sulphates is given in the German Tensid-Taschenbuch[7].

The comparison of surface tensions observed for the betaines mentioned herein yielded the following groups which are listed in figure 6 and illustrated in figure 7.

Table 2, Figure 8

7 REFERENCES

1. W.M. Linfield a.o., Tenside Surf. Det., 1977, 27, 3.
2. D.M. La Petina a.Ch. Patel, EP 253 489 dd. June 9, 1987.
3. H. Köbel a.o., Tenside Detergents, 1964, 1, 7.
4. G. Dardini a.o., Ann. Microbiol. Enzimol., 1966, 16, 217.
5. T.C. Cordon a.o., J.Am.Oil Chem.Soc., 1970, 47, 203.
6. P. Schöberl a.o., Tenside Surf. Det., 1988, 25, 86.
7. 'Tensid-Taschenbuch', 1990, p. 94 ff.

Applications

Diacid® — A Hydrotrope of Unusual Structure

Stig E. Friberg

CENTER FOR ADVANCED MATERIALS PROCESSING AND DEPARTMENT OF CHEMISTRY, CLARKSON UNIVERSITY, POTSDAM, NEW YORK 13699—5810, USA

A comparison is made between the basic mechanism of function of a traditional hydrotrope and a long chain variety, the Diacid®. Both kinds of hydrotropes function by disordering a lamellar liquid crystal causing a transition to an isotropic solution. Examples of this behavior are presented for concentrated solutions, mimicking liquid cleaner concentrates and the highly diluted conditions in the traditional laundry process. The disordering effect of the long chain compound is explained by its conformation at an interface as revealed by low angle X–ray diffraction analysis.

1 INTRODUCTION

The behavior of 5–carboxy–4–hexyl–2–cyclohexene–1–yl octanoic acid (Diacid®) as a hydrotrope is better understood after the mechanism of the traditional hydrotropes (1–5) has been explained. The main characteristics of these compounds (1,2) were outlined early by several researchers and may be summarized by the solubility curve in Fig. 1. It reveals the pronounced solubility increase of a hydrophobic substance, octanoic acid, in water, when the hydrotrope, sodium xylenesulfonate, is added to a level of 20% by weight, counted on the water content. The increase from fractions of one percent at low hydrotrope concentrations to approximately 90% is noteworthy and justifies an explanation.

The clarification was presented some time ago (6). It used a comparison between the solubilizing action on octanoic acid by a surfactant, sodium octanoate, and a hydrotrope, sodium xylenesulfonate, to outline the differences in structure during the solubilization process.

The difference in solubility, at a first glance, Figs. 1,2, may appear decisively in favor of the hydrotrope with 90% solubility versus only 10% for the

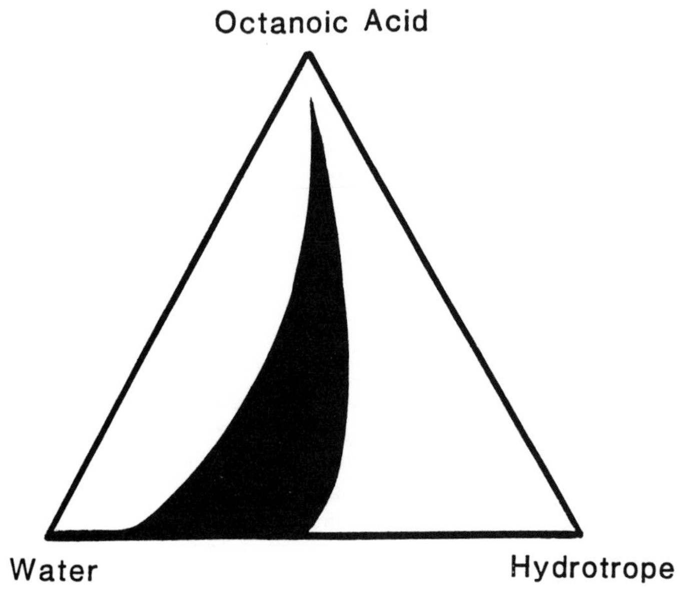

Figure 1 The solubility of octanoic acid in an aqueous solution of a typical hydrotrope, sodium xylenesulfonate, increases drastically after a high percentage of the hydrotrope (≈20% by weight) has been reached.

surfactant. However, these numbers only reflect the octanoic acid solubility in the aqueous isotropic normal micellar solution, the limit of which is determined by the lamellar liquid crystal. This later body is, of course, a separate phase from the aqueous solution, but its presence does in reality not mean a termination of the solubility of the octanoic acid. It is soluble in the liquid crystal as it is soluble in the inverse micellar solution. Hence, the solubility of octanoic acid in the aqueous solution of the soap should rather be viewed as represented by a single area as in Fig. 3.

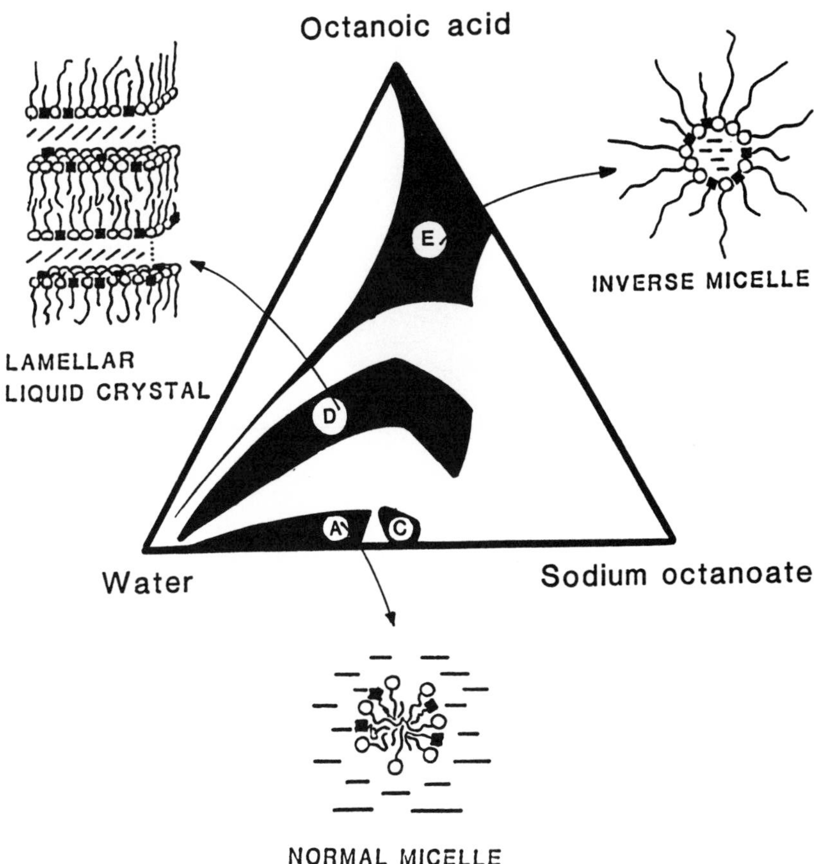

Figure 2 The solubility regions of octanoic acid in the water/surfactant (sodium octanoate) combination.

A Aqueous micellar solution
D Lamellar liquid crystal
C Liquid crystal of hexagonally packed cylinders
E Octanoic acid solution with inverse micelles.

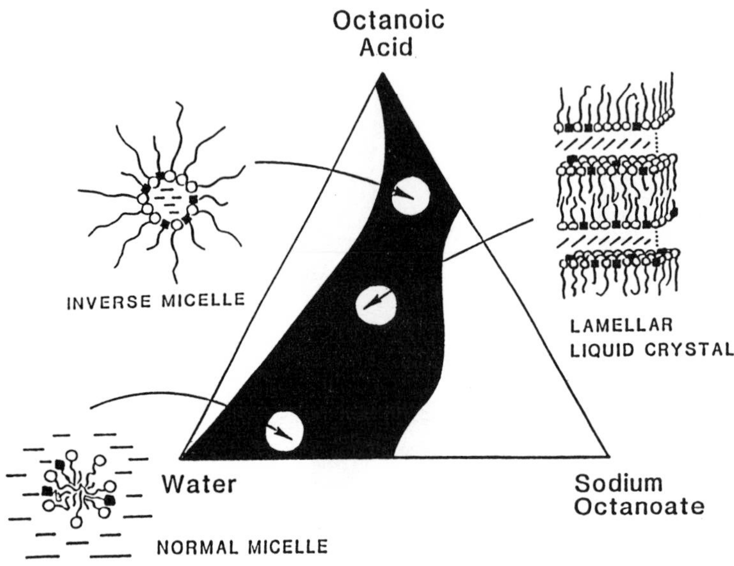

Figure 3 The total solubility region in Fig. 2.

The difference between the surfactant and the hydrotrope lies with the onset of solubilization and with the structure appearing during the transition from the water continuous system to the octanoic acid continuous one. The surfactant having a longer chain commences association at lower concentrations according to the well known relations (7). The second difference, the transition structure, is related to the interconnection between the compatibility of the amphiphile structure and the packing conditions to form a lamellar liquid crystalline phase.

The order of the hydrocarbon chains in a liquid crystal is described by the order parameter (8). It is directly obtained from the NMR quadrupolar splitting of the deuterium nucleus (9) and varies along the hydrocarbon chain according to Fig. 4. The order parameter remains approximately constant along the chain from the polar group and is strongly reduced towards the terminal methyl group. For equal chain length of the carboxylic acid and the surfactant the order parameter variation is identical, but unequal chain lengths mean that the part of the longer chain exceeding the shorter one will be highly disordered.

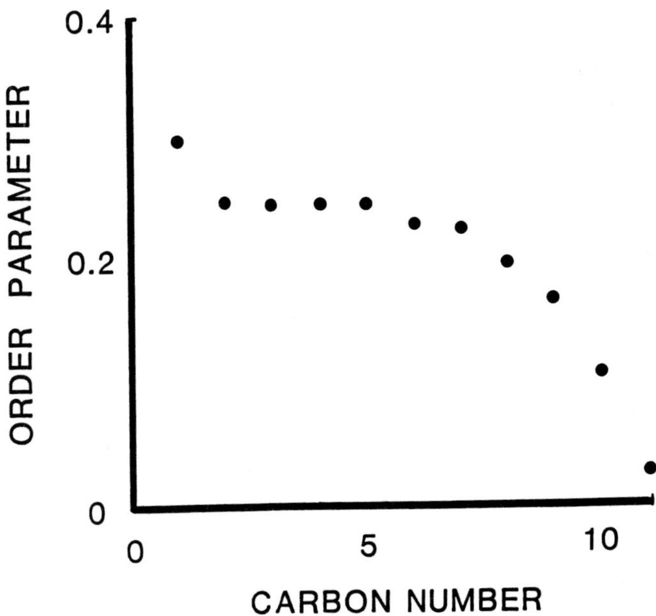

Figure 4 A typical variation of order parameters for the methylene groups of straight chain amphiphile in a lamellar liquid crystal. The carbon numbers are counted beginning with the methylene group adjacent to the polar group.

This latter effect is the key to the very high solubility of octanoic acid in the sodium xylenesulfonate solution, Fig. 1. The hydrotrope is ill matched to the octanoic acid molecule, Fig. 5, and the formation of a lamellar liquid crystal is prevented because the bulky hydrotrope structure, Fig. 5, will cause a pronounced disorder and, hence, instability of the liquid crystalline phase. The transition from water continuous association structures, Fig. 6A to organic (octanoic acid) continuous ones Fig. 6C does, hence, not take place through an ordered lamellar liquid crystal, Fig. 2, but through disordered structures, Fig. 6B.

The result is a continuous solubility region in the form of an isotropic liquid from the aqueous corner to the acid solution, Fig. 1.

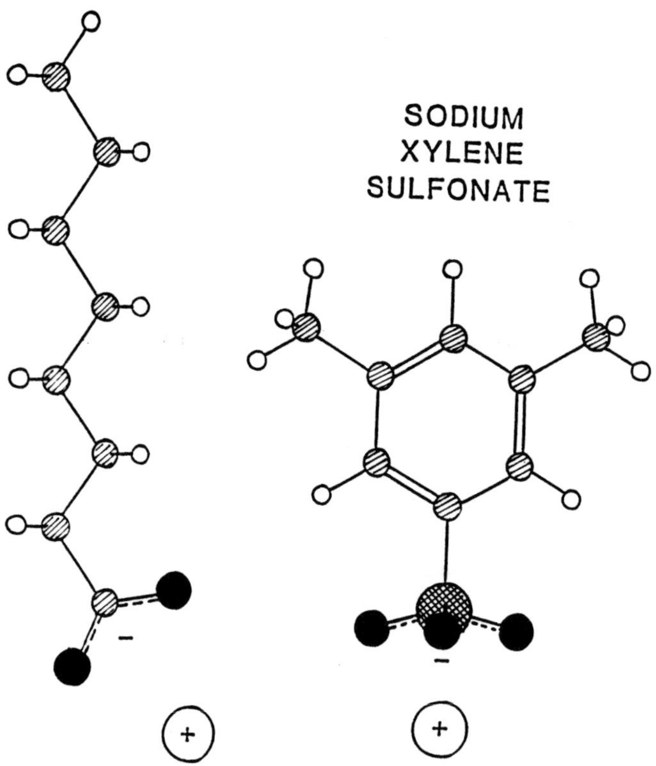

Figure 5 The structures of a surfactant, sodium octanoate, and a hydrotrope, sodium xylenesulfonate.

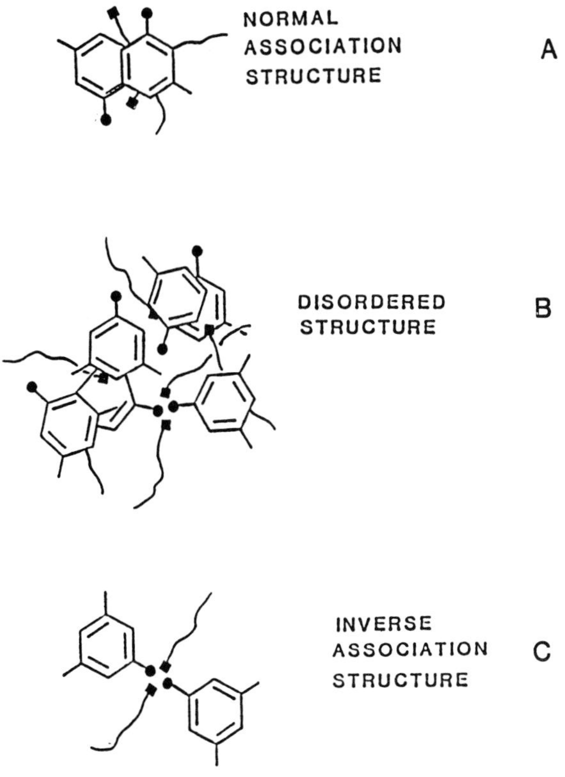

Figure 6 In the aqueous rich part, part of solubility region, Fig. 1, the association is between the hydrophobic parts of the octanoic acid and the hydrotrope, A. In the octanoic rich part, upper part Fig. 1., association takes place between the polar parts of the molecules, C. The transition region shows disordered structures, B.

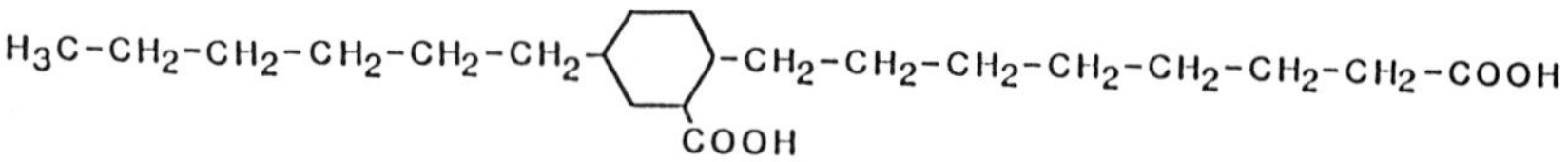

$H_3C-CH_2-CH_2-CH_2-CH_2-CH_2-$ [cyclohexene ring] $-CH_2-CH_2-CH_2-CH_2-CH_2-CH_2-CH_2-COOH$

COOH

Figure 7 The structure of 5–carboxy–4–hexyl–2–cyclohexene–1–yl octanoic acid, Diacid®.

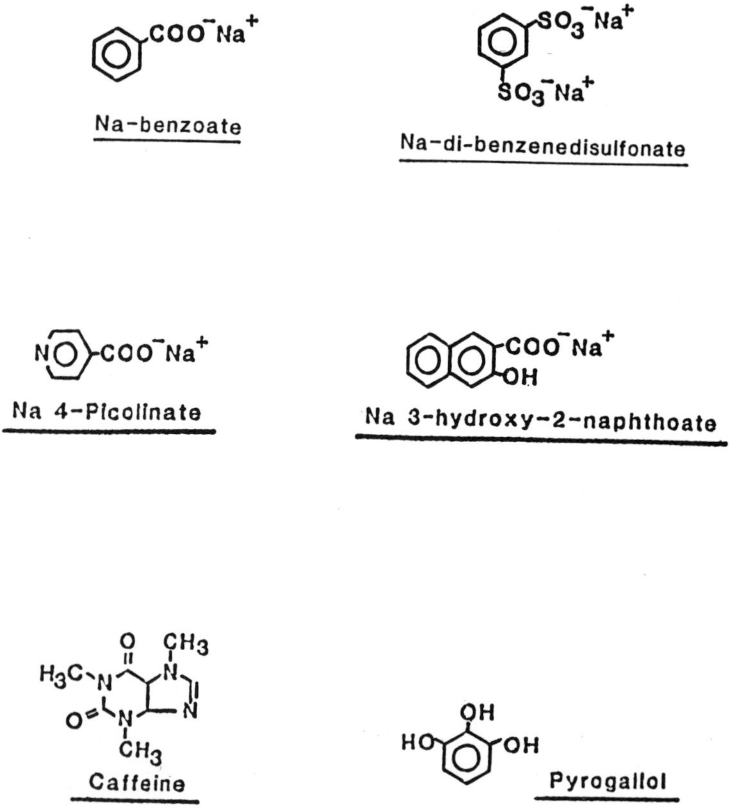

COO^-Na^+

Na-benzoate

$SO_3^-Na^+$

$SO_3^-Na^+$

Na-di-benzenedisulfonate

N COO^-Na^+

Na 4-Picolinate

COO^-Na^+

OH

Na 3-hydroxy-2-naphthoate

O CH3
H3C N N
O N N
CH3

Caffeine

OH
HO OH

Pyrogallol

Figure 8 Examples of traditional hydrotropes; all characterized by a short bulky structure.

The introduction of 5–carboxy–4–hexyl–2 cyclohexene–1–yl octanoic acid, Diacid®, by Westvaco (10) meant a hydrotrope of different structure, Fig. 7. Instead of the usual short bulky configurations of hydrotropes, Fig. 8, the Diacid® molecule contains two straight hydrocarbon chains connected by a hexenyl ring with a carboxylic group, Fig. 7. Its mechanism of hydrotropic action could not be expected to be as simple as that of the common hydrotropes and an examination of it is justified.

A Hydrotrope of Different Structure

The common application of hydrotropes in liquid cleaners requires two functions. The presence of the hydrotrope should prevent gelation in the concentrated formulations of the cleaner and it should prevent the formation of viscous gels, when the aqueous detergent solution contacts the "oily dirt" on the fabric.

The Diacid® hydrotrope (10) has been shown to achieve these goals. The action in a concentrated system, to imitate the liquid cleaner formulation, was illustrated with the combination of an amine (11). A common straight chain acid, octanoic acid caused gelation when combined with a amine amphiphile under water poor conditions, while combinations with the Diacid® only resulted in isotropic solutions.

In the same manner, it was demonstrated (11) that the lamellar liquid crystals formed, when a model "oily dirt" (12) was contacted by an aqueous solution of a surfactant, were changed to an isotropic solution when the Diacid® was added to the surfactant in appropriate amounts. These results showed the Diacid® to give the general response expected from a hydrotrope in action, but did not clarify the molecular details of its mechanism. These were revealed by investigations using NMR and low angle x–ray diffraction. The results of these demonstrated both the interaction of the Diacid® on the lamellar liquid crystal (13) and its conformation, which caused this interaction (14). The results are interesting to some extent because they revealed a new simple method to find the conformation of a complex molecule at an interface. It should, when adopted more generally, be a valuable complementary method to the common Langmuir Blodget technique (15).

NMR and LAXD Results

The preliminary results on the action of the Diacid® (11) revealed that its presence destabilized a liquid crystalline phase, but gave no indication as to the mechanism of destabilization. This query was answered by the NMR results (13).

These gave the change of order parameter for the "oily dirt" component when the Diacid® was added to the lamellar liquid crystal formed by the "oily dirt" model, the surfactant and the water. Fig. 9 clearly shows that addition of more surfactant did not change the order of the "oily dirt" component, while addition of the diacid hydrotrope led to a distinctive disordering of the entire molecule. As an

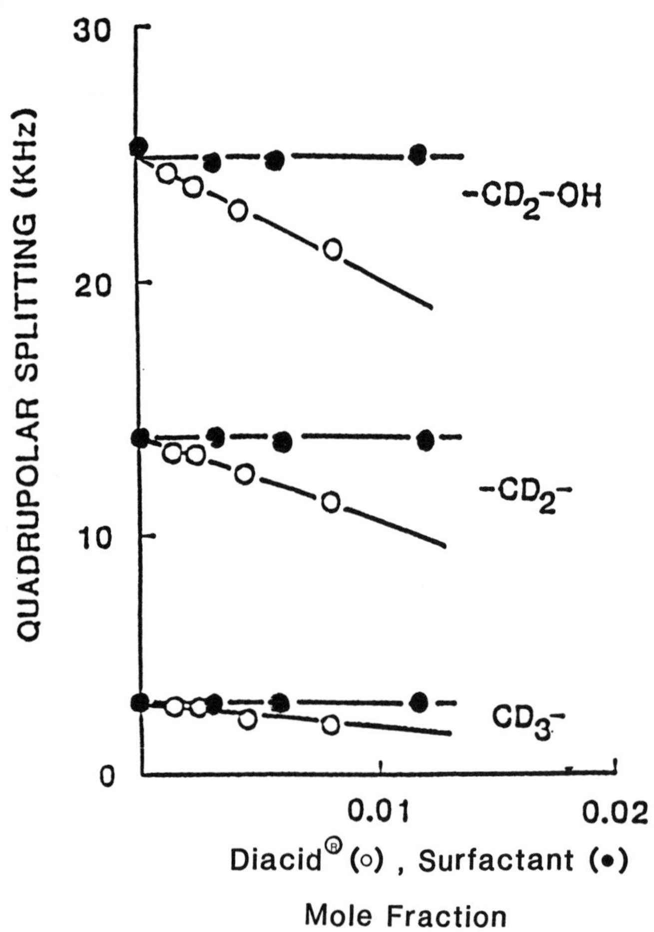

Figure 9 Addition of surfactant to a water/surfactant/model "oily dirt" liquid crystal (•) did not change the order parameter of the model "oily dirt" compound. Addition of Diacid® caused a reduction of order parameters along the entire chain (○).

illustrative example, may be mentioned a 20% reduction of the order parameter, when the hydrotrope addition was 10% by weight counted on the non–aqueous components.

The remaining question, the one of what causes the disordering effect on the liquid crystal, was analyzed using low angle x–ray diffraction (14) on the liquid crystal including the model "oily dirt". The x–ray diffractograms directly provide the interlayer spacing of the lamellar liquid crystal, Fig. 10A. The change of interlayer spacing, when an amphiphilic molecule is added, gives information about its reach into the lamellar structure. If it reaches less than the chain length of the original compounds the interlayer spacing will be reduced and vice versa.

A typical example is the model liquid crystal with "oily dirt" to which oleic acid was added. The presence of the oleic acid molecules caused an increase of the interlayer distance reflecting the conformation in Fig. 10B. The application to the problem of the Diacid® is summarized in Fig. 10C and D. If the conformation of the Diacid® were such that only one of the polar groups was anchored at the interface to water, Fig. 10C, the interlayer spacing would be increased after its addition to the liquid crystal with model "oily dirt". On the other hand, a conformation according to Fig. 10D with both polar groups attached to the water layer would mean no increase.

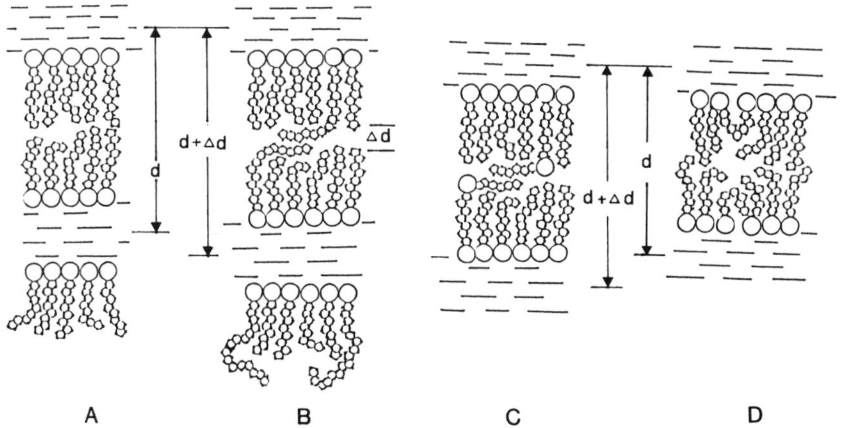

Figure 10 Adding an amphiphile with a longer chain to a lamellar liquid crystal (A) gives an increase of interlayer spacing (B). The two possible conformations of a Diacid® molecule within a lamellar liquid crystal (C and D) are distinguished by the interlayer spacing difference.

The results (14) gave an unequivocal answer. The interlayer spacing after addition of the Diacid® was unchanged, Fig. 11. This result means that the conformation in Fig. 10C is excluded; it would give rise to an increase of interlayer spacing similar to the one obtained after addition of oleic acid, Fig. 11.

The conformation of the Diacid® molecule is, instead, that of Fig. 10D with both the polar groups at the surface. This conformation provides an obvious explanation to the hydrotropic action of the Diacid®. The combination of the short loop and the attached short straight chain, Fig. 10D, is expected to cause pronounced disorder in a liquid crystalline phase. The NMR results, Fig. 9, show such a disordering to take place; Fig. 10D shows why so was the case.

The Diacid® has been successfully applied in several products. The following section deals with the results from some application oriented investigations.

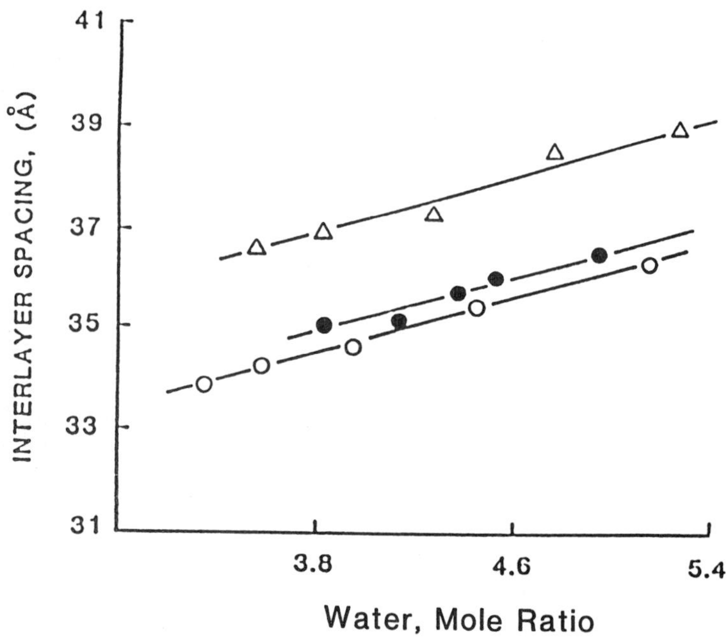

Figure 11 Addition of oleic acid to a water/surfactant/model "oily dirt" lamellar liquid crystal (○) gave the expected increase of interlayer spacing (△). No such increase was found after addition of Diacid® revealing its conformation to be that in Fig. 10D.

Application Oriented Research

The application of hydrotropes includes many areas outside liquid cleaners. The following review (18) covers results showing cloud point elevation for nonionic surfactants, increase of the practical maximum solubility of surfactants in water and improvement of the foaming of shampoos and syndet bars.

The cloud point of nonionic surfactants (ethylene oxide adducts) is defined as the temperature at which a dilute solution (1%–2% by weight) of the surfactant loses its transparency during temperature rise. The reason for this behavior is well known (16,17). The aqueous solution separates into two solutions; a lower consolute point has been reached, Fig. 12. Diacid® is more effective than the common short chain hydrotropes as shown by Table I (18).

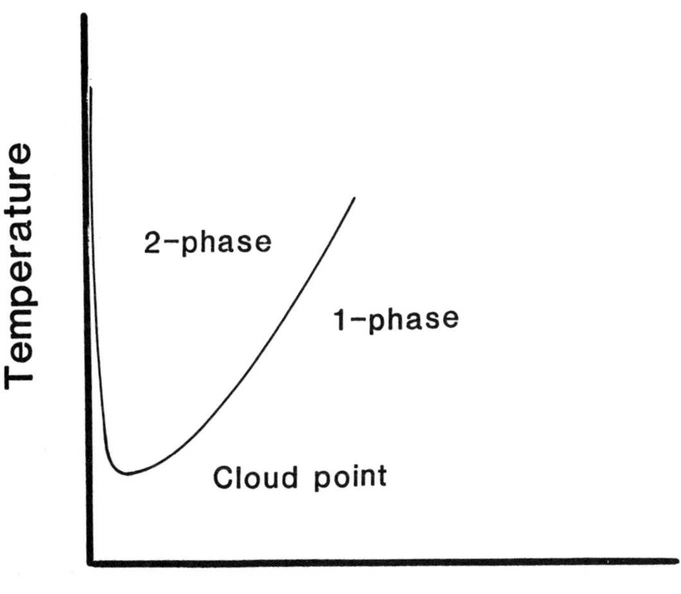

Figure 12 The cloud point in an aqueous solution of a nonionic surfactant (ethylene oxide adduct) is a lower consolute point.

TABLE I

The ratio between the average amount needed of Diacids® (1595, 1550) and the average of three common hydrotropes to form clear solutions.

Surfactant	Igepal co−530	Plurafae RA−40	Tetronic 701	Tergital 15−S−7	Neodol 23−6.5
Ra	0.5	0.1	0.1	0.7	0.9

The average Ra value is 0.5 clearly demonstrating the efficacy of the Diacid® for this purpose.

The enhanced maximum solubility of surfactants in the presence of the Diacid® is well illustrated by the cut off point for pourability of tall oil fatty soap (TOFS) solutions (18). The increase from 20 to 30 to 37% of the TOFS content with 0, 10, and 20% Diacid® (counted on the soap) expose the effect of the latter.

The high foaming and retained viscosity of shampoo formulations and syndet bars are other essential features of the application advantages of Diacid®

2 SUMMARY

A short review of the mechanism of hydrotropic action was presented and a comparison to the action of a hydrotrope with unusual structure, the Diacid®, was made.

The advantages were presented of this hydrotrope application in detergents, shampoos and syndet bars.

3 ACKNOWLEDGEMENT

The financial support of Westvaco Corporation for this research is gratefully acknowledged.

References

1. C.Neuberg, Biochem. Z. 76, 107 (1916).
2. H.Freundlich, G.V.Slottman, Biochem. Z. 188, 101 (1927).
3. F.-V.Hahn, Kolloid Z. 62, 202 (1933).
4. R.H.McKee, Ind. Eng. Chem. Ind. Ed. 38, 382 (1946).
5. P.H.Wilson, Trans. Faraday Soc. 34, 376 (1948).
6. S.E.Friberg and L.Rydhag, Tenside, 7, 80 (1970).
7. K. Shinoda in "Colloidal Surfactants" (K. Shinoda, ed.) Academic Press, New York (1963) p.1.
8. G.J.T.Tiddy, Nature (London) 230, 136 (1971).
9. J. Charvolin, P. Manneville and B.Deloche, Chem. Phys. Lett 23, 345 (1973).
10. B.F.Ward, C.G.Force, A.M.Bills, and F.E.Woodward, J. Amer. Chem. Soc. 52, 219 (1975).
11. J.M.Cox and S.E.Friberg, J. Am. Chem. Oil S. (1981).
12. H.S.Kielman, P.J.F.van Steen, "Surface Active Agents" Soc. Chem. Ind. London, (1979) p.191.
13. S.E.Friberg, S.B.Rananavare, and D.W.Osborne, J. Coll. I. Sci. 109, 2 (1986).
14. T.Flaim and S.E.Friberg, J. Coll. I. Sci. 97, 1 (1984).
15. N.L.Gershfeld and Y.G.Pak, J. Coll. I. Sci. 23, 215 (1967).
16. J.M.Corkill, J.F.Goodman and S.P.Harold, Trans. Faraday Soc. 60, 202 (1964).
17. K. Shinoda, J.Coll. I. Sci. 34, 278 (1970).
18. P.L.Robinson, J. Am. Oil Chem. Soc. (In Press).

The Use of Chemicals in Oilfield Demulsification

B. Rowan

PETROLITE LIMITED, KIRKBY BANK ROAD, KNOWSLEY INDUSTRIAL PARK (NORTH), LIVERPOOL L33 7SY, UK

This paper will give a general overview of the necessity to dehydrate crude oil in an efficient and cost effective manner. It will explain some of the theory of emulsions, factors affecting their stability and describe production systems used to resolve the immiscible liquids. The chemical composition of demulsifiers will be discussed together with the well accepted technique of selecting demulsifiers for oilfield application.

During the lifetime of an oilfield, water will be co-produced with the crude oil. This water may be already present in the formation or it may be there as a consequence of an enhanced oil recovery programme in which water is injected into the formation to maintain the reservoir pressure. It is uncertain when the emulsions are formed. There may be some sort of emulsion actually present in the formation or the emulsion may be formed as the fluids pass through the well tubing, valves, wellhead chokes, manifolds etc. A certain amount of shear associated with these restrictions to natural flow cause the immiscible liquids to become mixed together to form a stable dispersion. An emulsion can, therefore, be defined as the combination of two immiscible liquids, one dispersed in the other and stabilised by an emulsifying agent.

Generally, the water is dispersed in the oil. The oil is therefore referred to as the continuous phase and the water as the dispersed phase. The emulsifying agents can be any impurities that are present ranging from paraffin waxes, asphaltenes, resins and organic acids to inorganic particles, sands and silt. It may be the action of one or more of these impurities which causes the stabilisation of an emulsion.

Oil producers regard the resolution of these emulsions as essential and a lot of emphasis is put on the removal of most of the water in the emulsions. The pipeline operators regard the treatment as an important step to reduce corrosion and erosion of their lines. They impose certain specifications usually in the form of a certain level of Basic Sediment and Water (BS&W). Depending on the particular crude type and dehydration system, this specification may be less than 1% BS&W. These figures may not appear to be significant, however if an oilfield is producing 250,000 barrels (bbls) of oil per day and 1% is foreign matter, then the lines will carry 2,500 bbls p.d. of unsaleable material. Over the lifetime of the oilfield this can be regarded as a very significant loss of revenue. Together with this loss of revenue some corrosion and erosion is bound to occur which will require some kind of maintenance, again this reduces the profitability of the operation.

The refiners of the oil may also have specifications which they use. More commonly, they are concerned not only with the water present, but also the salt content associated with it. The removal of these impurities from the feedstock is regarded as an essential step in reducing the corrosion and fouling of refinery units and flowlines.

It is obviously economically beneficial for an effective oil dehydrating treatment programme to be employed.

1 THE NATURE OF OILFIELD EMULSIONS

As stated previously, the factors which lead to the formation of an oilfield emulsion are: two immiscible liquids, an emulsifying agent, and the energy required to physically disperse the liquids. The energy of mixing is supplied by the passage of crude oil through the flowlines, valves, chokes etc associated with the production system. This paper will only be concerned with the water-in-oil type of emulsion as opposed to the oil-in-water type or reverse emulsion.

There are many factors which affect the stability of an emulsion. Some oilfields produce emulsions which can be treated with as little as 2-4 ppm of chemical, on the other hand some emulsions will require 100-200 ppm of demulsifier to attain pipeline specifications. The factors affecting emulsion stability include:-

Emulsifying Agents

Emulsifiers are surface active agents or solids which concentrate at the oil/water interface. Their ability to form stable emulsions will depend on their speed of migration to the interface and their performance when they arrive at that site. The role of an emulsifier is to reduce the surface tension and form a physical barrier around the droplet. Over time, the emulsifier produces a thicker and tougher film around the droplets thereby forming a more stable emulsion.

Agitation/Water Droplet Size

The size of the dispersed water droplet is determined by the degree of agitation in the system. The more severe the agitation the more shearing action is imparted, and consequently smaller water droplets are produced, and a more stable emulsion results.

Water Content of the Oil

Generally as the water content of the oil increases the stability of the emulsion decreases. This is related to the lower concentration of emulsifiers at the oil/water interface and the larger average size of the water droplets in the oil.

Viscosity of the Oil

The viscosity of the oil phase plays a dual role. In a high viscosity oil, a given amount of agitation will not break up the water droplets as much, as would be the case with a lower viscosity oil. However, a high viscosity oil is able to maintain larger water droplets in suspension and smaller droplets will have a greater resistance to settling.

In general higher viscosity oils are more difficult to dehydrate than lower viscosity oils.

Specific Gravity

The greater the difference in specific gravity between the oil and water phases the faster the water can settle. Higher API gravity oils (lighter oils) can be dehydrated easier than heavier crudes.

Temperature

An increase in temperature enables emulsions to be resolved more easily.

An increase in temperature will decrease the viscosity of an oil thereby allowing easier coalescence of the water droplets. Secondly, by increasing the temperature the emulsifier film surrounding the water droplet is weakened, and thirdly the larger the difference in specific gravity between the fluids the larger the tendency to separate under gravity.

Ageing of Emulsions

Since emulsifiers are surface active, they migrate to the oil/water interfaces thereby stabilising the emulsion. As time progresses more emulsifiers migrate to the interface forming a thicker film and making the emulsion more difficult to break.

2 THE NATURE OF DEMULSIFIERS

The generally accepted method of resolving oilfield emulsions is by the application of a suitable demulsifier. These products are a blend of highly surface active intermediates normally based in an organic solvent. For the demulsifier to be successful it must fulfil the following criteria:-

Attraction of the Oil/Water Interface

For any separation to occur the demulsifier must migrate to the oil/water interface. When the demulsifier is at this site it must then compete with the emulsifier which will be surrounding the water droplets.

Flocculation

When the demulsifier is located at the surface of the water droplet, the droplet has a strong attraction for other droplets in the same condition. Therefore large agglomerations of water droplets are produced. This flocculation process may not be sufficient to allow the emulsion to be resolved. The emulsifier film may remain strong enough to prevent any resolution of emulsion.

Coalescence

When the emulsifier film is totally ruptured and the water droplets commingle, this state is known as coalescence. As the complete disruption of the emulsifier film occurs, the growth in water droplet size increases and rapid separation of water results.

Solids Wetting

In most crude oils, solids such as Iron Sulphide, silt, clay, drilling mud etc, complicate the demulsification process. They tend to collect at the interface and contribute significantly to emulsion stability. Sometimes such solids are the primary stabilising material and their removal is all that is necessary to achieve satisfactory emulsion resolution. For removal from the interface, these solids can be dispersed in the oil phase or water wetted and be removed with the water. If they are dispersed in the oil they will remain as a contaminant in the oil. It is, therefore, generally desirable for any inorganic solids to be removed with the water. Paraffin and other organic solids are the exception to this, since they can be recovered in the refining processes it is more beneficial to remove them with the oil.

There are, therefore, four primary actions of a demulsifier necessary to resolve the emulsion produced in an oilfield. As shown there are many different factors affecting the stability of the emulsion. It is not surprising that one single chemical structure cannot be effective in fulfilling all these actions. More commonly, the most cost effective demulsifier will be a blend of two or three chemical structures.

As has been already stated the most cost effective chemical for treating an oilfield emulsion is normally a blend of 2 or 3 chemical types. The chemicals tend to fall into one of three categories, as follows:-

a) Oxyalkylated Phenol/Formaldehyde Resins
This group can be subdivided depending on how the resin is prepared.

   i) Acid Catalysed Resins

   ii) Base Catalysed Resins

b) Polyglycols

c) Miscellaneous - Oxyalkylated Phenols and Alcohols
Oxyalkylated Amines
Sulphonates
Esters

Acid Catalysed Resin. These are straight chain alkyl phenol formaldehyde. These resins are then oxyalkylated using ethylene oxide which reacts with the OH groups forming poly ether chains.

Base Catalysed Resin. The main difference between the base catalysed and acid catalysed reaction is that the base catalysed resins are more branched, and contain more cyclic material.

Derivatives of Polyoxypropylene Glycols. These products involve the propoxylation of a glycol followed by its reaction with an epoxy resin. The resulting diepoxide can be either reacted with propylene or ethylene oxide to produce various demulsifying intermediates.

Miscellaneous. As stated, this includes polyamines polyglycol esters, sulphonates, oxyalkylated phenols and alcohols and are generally used as minor components in conjunction with the other main groups of demulsifiers.

While some of the demulsifier intermediate types described above, in some instances, give complete demulsification when used alone, the performance can usually be enhanced by blending intermediates from different groups to produce the most cost effective chemical for a particular application. Selection of the best formulation takes experience and requires extensive training. Meaningful demulsifier testwork must always be carried out in the field on fresh samples of crude oil. One is unlikely to select the optimum product with work performed on "aged" or "dead" crude samples. The procedure for selecting the demulsifier is by a series of laboratory tests which is known as the "Bottle Test".

3 METHOD OF SELECTING DEMULSIFIERS

The Bottle Test can only be regarded as a comparative test. That is, it will only select the most effective chemical from a series of tests under carefully controlled conditions. It will not give the actual dosage rates required, although the system dosage rate is normally less than the dosage required in the bottle.

Before embarking on the bottle test it is essential to have a full knowledge of the production system. Temperatures, residence times in vessels and site of chemical injection are all parameters which are required so that the bottle test is performed to simulate the system as closely as possible. This will then give the chemical which is selected the best possible chance of working in the system.

Consider, for example, a very simple production system. Crude oil flows from the wellhead to a manifold, demulsifier is injected and the fluids flow to a single 3 phase separator operating at 60°C. The residence time in this vessel is 20 minutes, after which time the oil outlet from the vessel contains 0.5% BS&W. The following procedure could be adopted to select a chemical.

1) An untreated emulsion sample is taken.
2) The sample is poured in 100 ml demulsification tubes and heated to 60°C.
3) Chemical is injected into each tube leaving one blank.
4) The tubes are shaken to disperse the chemical and placed back in the water bath at 60°C.
5) The water separated is noted periodically.
6) After 20 minutes, the tubes are removed and a sample from the top oil is taken and the BS&W determined.

Obviously any chemicals that give fast and complete separation of water together with a top oil BS&W of less than 0.5% within 20 minutes would be regarded as effective chemicals for this particular application. It is important to realise that the criteria for chemical selection depend on the production system. If the residence time in the system is 24 hours, it would be pointless to use a chemical which will separate 90% of the water in the first hour and only another 5% in the remaining 23 hours. In this case speed is not required whereas complete dehydration is. If we consider the other extreme, on some offshore installations the residence time may be as short as 2 or 3 minutes. Obviously, speed of dehydration becomes a major factor in product selection in this case.

As a general rule, the BS&W level achieved in the bottle test must match or be better than the customers shipping specification.

4 CHEMICAL APPLICATION AND PRODUCTION SYSTEMS

In order to obtain maximum benefit from the chemical, the best site for chemical injection must be chosen. It should be located at a point far enough upstream from the treatment vessels to allow adequate dispersion. A good point is at the wellhead or manifold, or at the site of some restriction e.g. valve, choke etc., where there should be enough turbulence and sufficient time for the chemical to perform its function on the emulsion.

The amount of chemical required to totally dehydrate the oil can vary from field to field. In general a lighter crude oil (such as those found in North Sea) is easier to treat than a heavier oil. This is a function of the viscosity of the crude oil and is due to the fact that for a given amount of agitation the chemical will disperse in the light oil more easily than the heavier oil. Because of the many factors affecting emulsion stability, the treating system and specifications defined by the operator, the dosage rates can vary from 2 ppm in the North Sea to 100+ ppm for a heavy crude.

There are many different types of oil production systems in use throughout the world. They can vary from a small pumping unit producing 10 barrels of oil per day, or a large producing well in the Middle East making 10,000 barrels of oil per day.

As expected, the associated system can also vary. Offshore installations, where space is at a premium, generally have either a series of Free Water Knockouts or 3 phase separators. The residence time in these systems tends to be short (between 2 mins and 1 hour) depending on the throughput.

Producing areas onshore may have the same kind of system of separators but more likely is the presence of wash tanks and electrical dehydrators. Where they have the space available they can install a system which will have a longer residence time. Consequently, the quality of the dehydrated crude should be very good and is often sold on a salt specification rather than BS&W. Some oilfields can obtain salt contents in the region of 20 ppm salt or less. Some of the more common dehydrating vessels are discussed below:-

<u>Free Water Knockouts</u>

In general, free water knockouts are used to separate large amounts of free water ahead of the treating plant. Free water is defined as any water that usually separates within a very short time period (5 minutes) when the fluids are stationary. The object is to remove all the free water so that only the emulsion is handled by the treating system. Free water knockouts are generally used in oilfields which have a high water to oil ratio.

Three-Phase Separators

These vessels are very common on offshore installations and the system would normally be arranged with other separators in series. As their name suggests gas, oil/emulsion and water are all separated. The water would normally flow to some kind of treating process where any residual oil is removed giving a better quality water for disposal. The oil/emulsion phase would then flow to another separator for further processing. Residence time in these vessels if used offshore is of the order of a few minutes.

Settling Tanks

Tank dehydration would be likely to be found in an onshore oilfield. The residence time in these vessels is usually measured in days since they have a high volume to throughput ratio. The speed of water separation is not, therefore, regarded to be of paramount importance. Because of the long residence time the quality of the dehydrated crude is usually very good. BS&W specifications in the order of 0.1% or less, can be obtained with the correct demulsifier.

Variations of settling tanks are often found on onshore production sites. One such example is the Wash Tank - this vessel is designed so that the inlet is situated at the bottom of the tank and the fluids pass up through the water layer and an interface layer. The interface layer must stabilise at some acceptable thickness. This interface can act as a filter for solids and unresolved emulsion and can, therefore, help the treatment process.

Electrical Dehydration

Electrostatic treaters (often called Chemelectric Units) operate in a similar fashion to other horizontal treating vessels except that high voltage electric grids are arranged along the length of the vessel. The electric field promotes excellent water coalescence by inducing a small charge on either side of the water droplets. When charged, the water droplets are attracted to each other with enough force to cause the formation of larger droplets which then settle out.

Because of the action of the electric field the chemical dosage can usually be decreased to effect the same treatment of the oil. These Chemelectric Units can be located both onshore and offshore and are frequently placed at the end of the treating system. With correct maintenance and chemical injection, good quality oil is produced, thereby increasing the revenue generated.

5 SUMMARY

The use of chemicals to resolve oilfield emulsions is the primary technique used by oil producers to provide saleable oil to refineries. Because of the dosages injected they can be regarded as the most cost effective method of separating water from oil.

The Chemistry and Uses of Food Grade Emulsifiers

J.A. Wellard and A.J. Davidson

HARCROS CHEMICALS PTY LTD., P.O. BOX 232, SEVEN HILLS, NEW SOUTH WALES 2147, AUSTRALIA

INTRODUCTION

The scope of this Paper is to discuss the range of food emulsifiers described as fatty acid esters, their chemistry, synthesis and application, as well as the legal framework which regulates their use. To be complete any review of food emulsifiers would, of necessity, include a wider range of natural and modified products which are used functionally as emulsifiers by the food industry. Such a list would include functional proteins from both animal and vegetable sources, natural colloids from marine and vegetable sources, as well as fatty acid esters, which are the focus of this Paper.

Increasingly consumers have demanded of the food industry products of greater variety and convenience, with superior quality and extended shelf-life. This trend has developed with a background of increased industry regulation and consumer awareness, particularly in regard to labelling requirements.

While, from a functional viewpoint, the term "food emulsifiers" covers a broad range of both "natural" and "synthetic" compounds, there is little doubt the consumers' perception is that we are discussing products of the chemical industry. The terms "chemical" and "additive" tend to be used synonomously.

That fatty acid esters are made in chemical plants by chemical processes is not in dispute, although biotech synthesis is an increasing possibility. For the synthesis of fatty acid esters, however, the chemical industry uses as its feedstock essentially naturally occurring raw materials, refined and modified, to produce not only functionally superior emulsifiers, but also products meeting specific dietary, even religious, requirements.

Although not all countries are uniform with regard to the fatty acid ester food emulsifiers they will permit, there is broad acceptance of the following categories reviewed in this Paper:

- Mono and diglycerides of fatty acids.
- Sorbitol esters of fatty acids and their ethoxylates.
- Diacetyl tartaric esters of fatty acid monoglycerides.
- Fatty acid lactylates.
- Polyglycerol esters of fatty acids.
- Propylene glycol esters of fatty acids.
- Sucrose esters of fatty acids.
- Acetic, lactic, citric and tartaric esters of fatty acid monoglycerides.

In the food industry, the term emulsifier is used synonomously with the range of fatty acid esters reviewed in this Paper. While it is true that some applications approximate to oil/water [ice cream] or water/oil [margarine] emulsions, our contention is that "food surfactants" would more appropriately describe the range of functional properties and commercial applications we will consider.

Finally, the Regulatory Status of food emulsifiers and their usage will be described as there is a rigid, structured system governing the approval and use of food additives.

PRODUCTS & THEIR CHEMISTRY

(i) Mono- & Diglycerides

Commercial products are mono- and diglyceride mixtures of varying proportions. The monoglyceride exists predominantly in the alpha form, with the beta form accounting for 10 - 15% of the total. Although it is possible to synthesize monoglycerides directly from fatty acid and glycerol, the more economical process is transesterification of triglycerides using an excess of glycerol. [Figure 1]

This process is conducted under alkaline conditions cleaving a fatty acid from the triglyceride which is accepted by a free hydroxyl in glycerol to form a new ester linkage. This is an equilibrium process and is used to produce typically 40% monoglycerides. By increasing the glycerol "excess" it is possible to obtain 60% monoglycerides, however, glycerol recovery is an important consideration.

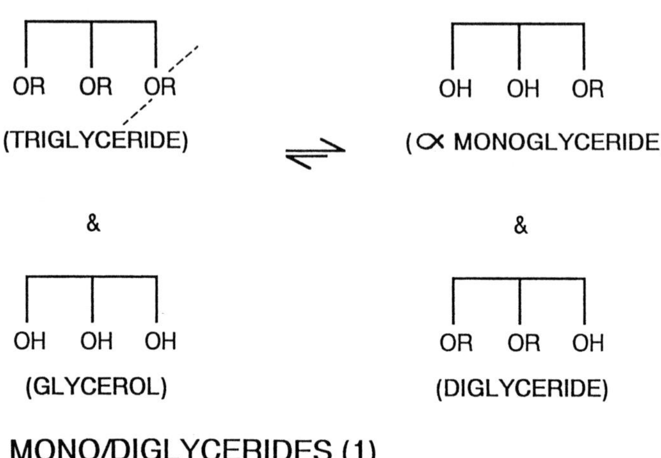
OR OR OR
(TRIGLYCERIDE)
&
OH OH OH
(GLYCEROL)
OH OH OR
(∝ MONOGLYCERIDE)
&
OR OR OH
(DIGLYCERIDE)
MONO/DIGLYCERIDES (1)

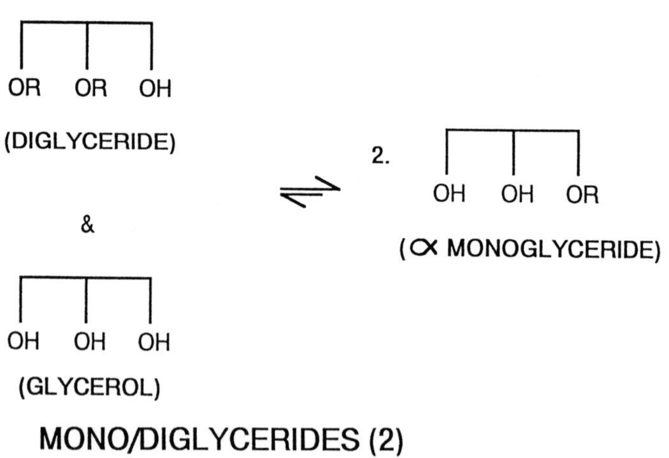
OR OR OH
(DIGLYCERIDE)
&
OH OH OH
(GLYCEROL)
2.
OH OH OR
(∝ MONOGLYCERIDE)
MONO/DIGLYCERIDES (2)

Figure 1

Commercial products, to 60% monoglyceride, are produced with this technology; the higher 90% monoglycerides by molecular distillation of lower mono- material. Biotechnology, in the form of enzyme catalysts, may provide an alternate approach to higher monoglycerides.

This technology is readily transferable, being applicable to a wide range of trigylcerides. Triglycerides may then be chosen on the basis of:

- desired functional or physical properties, e.g. HLB value [ice cream is an oil/water emulsion, margarine a water/oil emulsion]

- organoleptic considerations [taste]

- dietary [vegetable or animal]

- religious considerations, Kosher or Halal.

These considerations aside, monoglycerides are defined in terms of the following chemical criteria:

- Monoglyceride level [alpha and/or total] e.g. for dough "anti-staling" a high monoglyceride level is required.
- Free fatty acid [FFA].
- Iodine value [IV].
- Free glycerol.
- Saponification value.
- Soap content, self-emulsifying [SE] or [NSE].

It is generally assumed that emulsifying, or functional, efficiency is directly proportional to alpha monoglyceride level. In a strict emulsifying application, e.g. margarine, this may be so, but experience suggests other factors are also relevant. Two example are ice cream and peanut butter.

With ice cream GMS [glyceryl mono-stearate] is added to ensure effective and stable dispersion of fat globules through the product. Other factors influenced are texture and smoothness with the commonly held view that the mono/di ratio in 60% product is more relevant for overall performance. For peanut butter GMS is added to prevent oil separation and enhance mouth feel. In this instance crystallizing temperature of the emulsifier is a most important consideration.

Monoglycerides can be considered perhaps the original and still most widely used, of the fatty acid ester type food emulsifiers. Other food emulsifiers use monoglycerides as an intermediate in their manufacture, or have been developed to improve functional properties.

Monoglyceride technology is hence basic to the study of ester food emulsifiers and its products still find wide application in the following food sectors :

- Ice Cream
- Margarine
- Baking
- Confectionery
- Peanut Butter

<u>(ii) Sorbitol Esters & Ethoxylates</u>

Monoglycerides are fatty acid esters based on the polyhydric alcohol glycerol. The investigation of esters based on other naturally derived polyhydric alcohols, e.g. Sorbitol, was a natural development. The commercial products on the market today are mono- and triesters usually of stearic and oleic acids, also their 20 : 1 ethoxylates designated Polysorbates.

Perhaps the most commercially interesting derivatives are the sorbitan monoesters. Their chemistry involves two processes; esterification of hydroxyl with selected fatty acids and 1-4 ring closure by etherification of the sorbitol. This is shown simplistically in Figure 2, although industrial practice

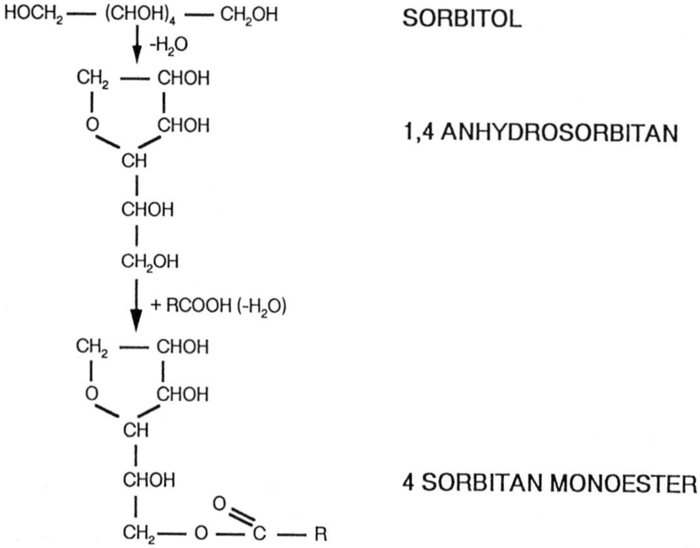

$HOCH_2—(CHOH)_4—CH_2OH$ SORBITOL

↓ $-H_2O$

CH_2 — CHOH
O CHOH
CH
CHOH
CH_2OH

1,4 ANHYDROSORBITAN

↓ $+ RCOOH (-H_2O)$

CH_2 — CHOH
O CHOH
CH
CHOH
$CH_2—O—C(=O)—R$

4 SORBITAN MONOESTER

<u>Figure 2</u> SORBITOL ESTERS

involves selection of catalyst and process such that both reactions occur simultaneously.

Ring closure of the sorbitol between the 1,5 positions also occurs and Figure 3 envisages a wider range of the reaction possibilities. Commercial sorbitan monoesters will contain both diester and sorbide monoester, while reaction conditions favourable to sorbitan triester formation likely yield more sorbitol and less sorbide monoesters.

Ethoxylation of sorbitan mono- and triesters [20 : 1 EO ratio] increases their hydrophilic character such that the formulator is offered a range of emulsifiers covering a wide HLB range.

Although these compounds demonstrate functional properties in a range of foodstuffs, for the most part food technologists use sorbitan esters and ethoxylates as co- or secondary emulsifiers. Polysorbates are used in conjunction with other emulsifiers in bread fats and industrial margarines. Being hydrophilic, they synergistically influence the functional properties of the more hydrophobic emulsifiers. Likewise both esters and ethoxylates are used in ice cream with GMS to improve heat shock and texture. Sorbitan monostearate is used synergistically in a variety of baked goods and non-dairy coffee whiteners.

Sorbitan monostearate finds application as the principal surfactant in yeast rewetting systems, while sorbitan tristearate is an effective anti-bloom agent in compounded chocolate. With the use of cocoa butter substitutes there is a greater propensity for fat migration to the chocolate surface, called "bloom", which the use of sorbitan tristearate [STS] retards by modifying the fat crystals.

(iii) Diacetyl Tartaric Esters of Fatty Acid Monoglycerides

In the commercial synthesis of these compounds diacetyl tartaric anhydride is prepared as an intermediate using acetic anhydride and tartaric acid. Esterification then proceeds with a selected fatty acid monoglyceride. [Figures 4 (a) & (b)]

While stoichiometric requirements for diacetyl tartaric anhydride formation are satisfied by a 1 : 1 ratio acetic anhydride to tartaric acid, in practice excess anhydride, with acetic acid formation, is used to expel the water produced and control formation of the monoacetyl derivative. The ratio of diacetyl to monoacetyl tartaric anhydride controls the tartaric acid level in the final product; a major determinant of functionability.

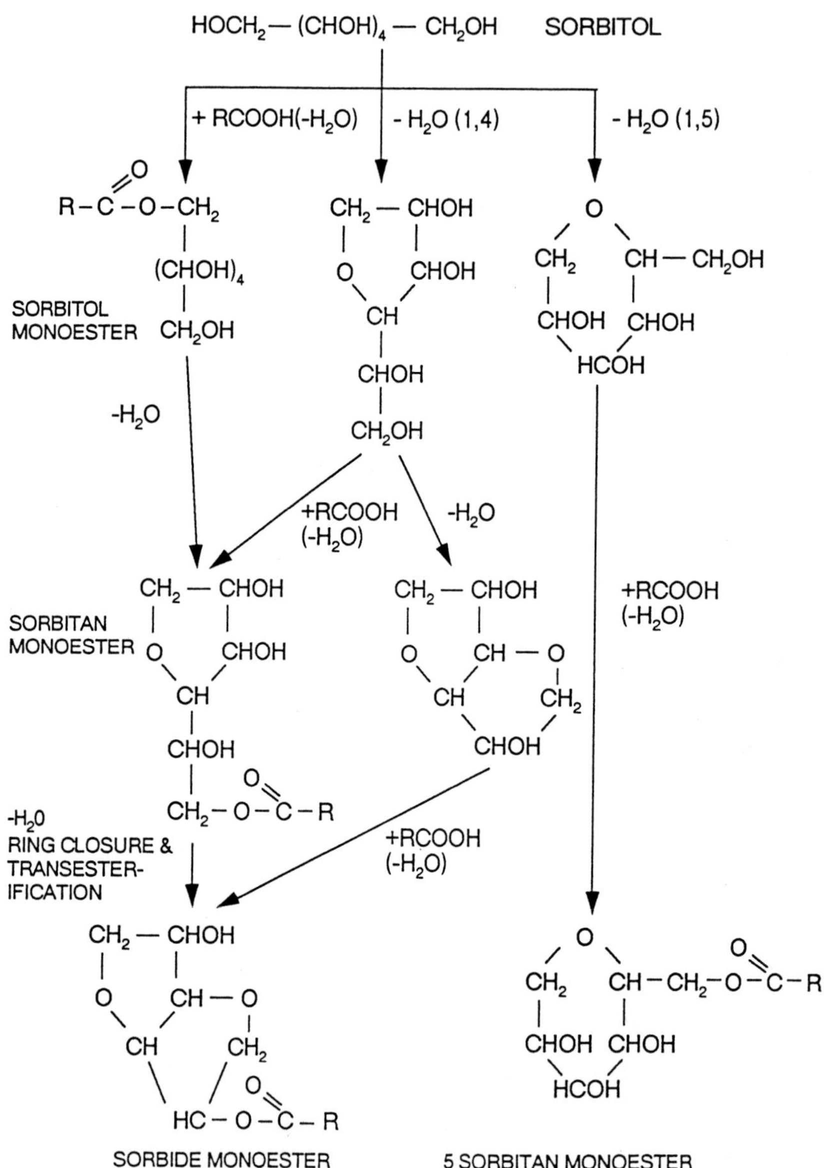

Figure 3 SORIBITOL ESTERS (b)

3.

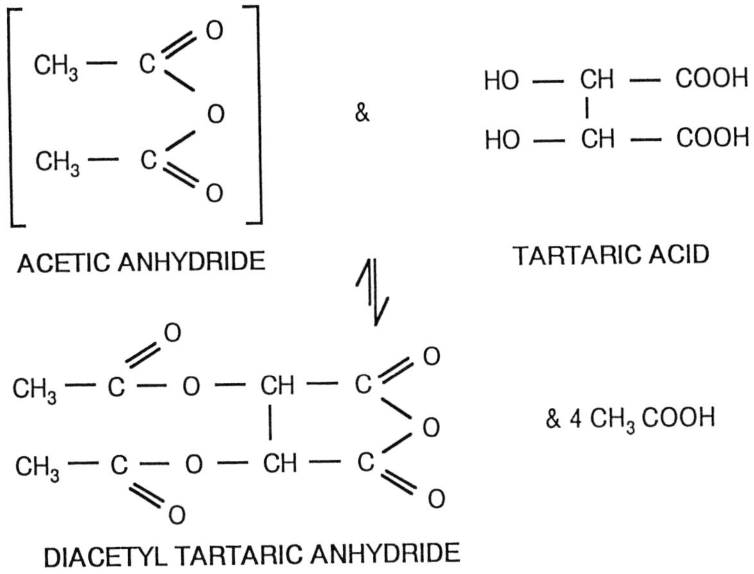

$[CH_3-C(=O)-O-C(=O)-CH_3]$ & $HO-CH(COOH)-CH(COOH)-OH$

ACETIC ANHYDRIDE TARTARIC ACID

⇌

$CH_3-C(=O)-O-CH-C(=O)-O-C(=O)-CH-O-C(=O)-CH_3$ (cyclic) & 4 CH_3COOH

DIACETYL TARTARIC ANHYDRIDE

Figure 4(a) DATA ESTERS OF MONOGLYCERIDES

DIACETYL TARTARIC ANHYDRIDE

⇌ & OH OH OR

$CH_3-C(=O)-O-CH(-C(=O)OH)-CH(-C(=O)-O-)-O-C(=O)-CH_3$ HO RO

DATA ESTER OF FATTY ACID MONOGLYCERIDE

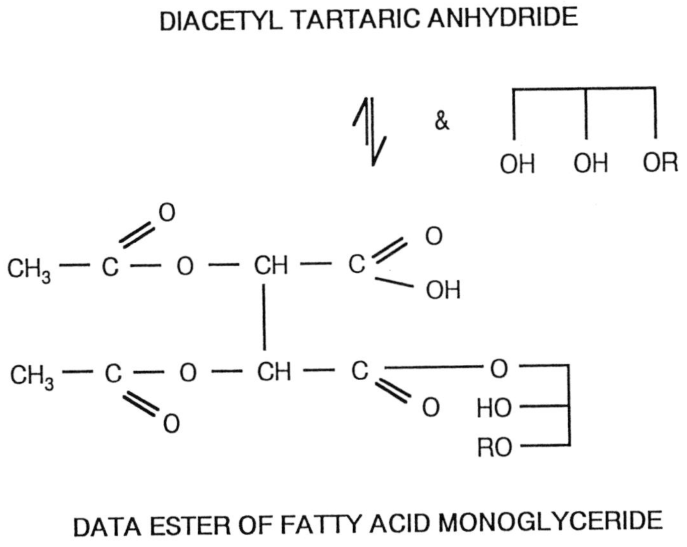

Figure 4(b) DATA ESTERS OF MONOGLYCERIDES

The other major determinant of functionability is choice of monoglyceride, particularly with regard to monoester and IV, since all commercial products are based on C_{16}/C_{18} fatty acids. Monoglycerides used are normally 60-90%, while IV covers the range fully saturated through GMO [IV90].

Almost exclusively, the DATA esters are used in bread and baked goods. Compared to monoglycerides DATA esters have lower starch [amylose] complexing efficiency [anti-staling], but being more water soluble draw both the emulsifier system and accompanying fat into the dough more quickly. DATA esters, by interacting with wheat protein [gluten], have a dough conditioning effect which strengthens structure, improves gas retention and volume, while enhancing crumb, softness and shelf-life.

The potential range of products described by this technology is large, but the formulators' final choice will be determined by method of use and the particular end result desired. If added directly to flour, or via a dry mix bread improver, a powder emulsifier is required. These are either milled or spray cooled, often incorporating an anti-caking agent, since they are low melting and quite hydroscopic. The same DATA ester active may be suitable for incorporation into a bread fat, although more plastic products [higher IV] are also used. For so-called pumpable shortenings even more liquid DATA esters are available.

(iv) Fatty Acid Lactylates

It is a feature of lactic acid, alpha hydroxy propionic acid, that it has both acid and alcohol functionability. Under controlled conditions inter-esterification occurs to form lactylic acid, which is capable of further esterification with fatty acids. [Figure 5]

Commercially, to meet Food Chemical Codex, the mole ratio lactic : stearic is less than 2 : 1, so salts of both acids will also be present in fatty acid lactylates. The most important fatty acid lactylate is SSL, sodium stearoyl - 2 - lactylate. Being water dispersable [SSL is quite hydroscopic] and combining high amylose complexing efficiency with protein interaction, SSL is widely used for bread anti-staling. By contrast with the sodium salt, CSL [the calcium salt] is less water but more oil soluble. Incorporation into the dough is thus recommended through a bread fat.

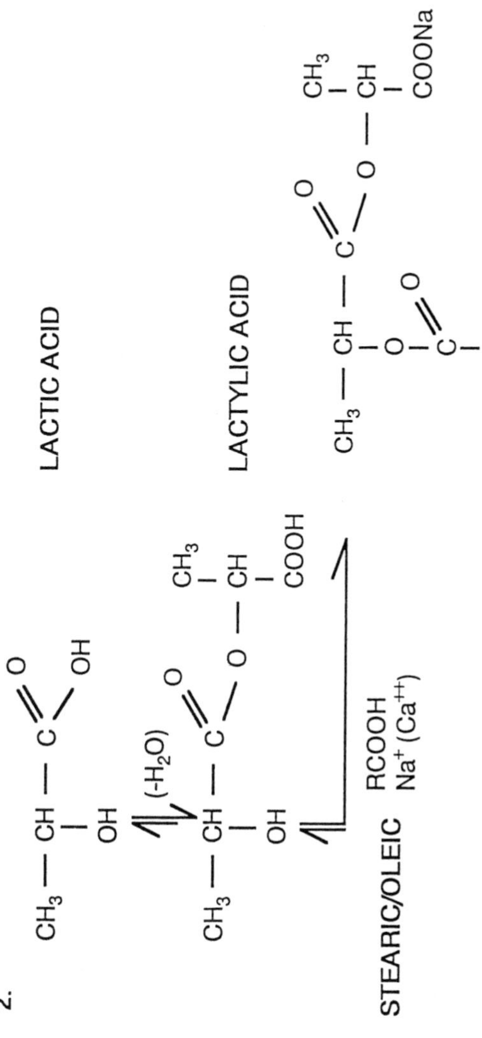

Figure 5 FATTY ACID LACTYLATES

Being both anionic and hydrophilic, SSL also may be used as a secondary emulsifier with more hydrophobic products, such as polyglycerol and propylene glycol esters. Indeed it is sometimes necessary to "activate" the functionability of such emulsifiers.

(v) Polyglycerol Esters of Fatty Acids

Of all the fatty acid ester categories, those of polyglyerol present most possibilities of synthesis. Despite the vast number of moities possible, relatively few have been commercialized and even now this category would probably present less than 5% of the total market.

It is not within this Paper's scope to discuss their chemistry comprehensively; a detailed review being available in McIntyre's Paper JAOCS, 1979 (1). Perhaps it is sufficient to canvas the possibilities in terms of:

- Polyglycerol; degree and type of polymerization.

- Ester; choice of fatty acid and degree of esterification.

Commercially polyglycerol is prepared as an intermediate, then esterified with fatty acid. An alternate approach is transesterification of polyglycerol with triglyceride. Concerning direct esterification, it is generally considered that glycerol, with alkali around 200°C, polymerizes linearly forming ether linkages between adjacent alpha hydroxy groups. Cyclic and mixed cyclic/linear polyglyerols are also possible and, compared to linear, reduce the number of hydroxyl groups available for esterification. Studies of OHV during and after glycerol condensation suggest the presence of non-linear polymer, as does the occasional report of a "set" during polyglycerol production. Figure 6 demonstrates the formation of linear polyglyercol, together with a Table showing OHV and free hydroxyl groups in linear polyglyerols.

The choice of fatty acid and degree of esterification will affect both physical and functional properties; oleic esters, for example, are more likely to be liquid products. A broad range of HLB products can be envisaged with hydrophilic surfactants at the low Dp [glycerol], low degree of esterification end, through more hydrophic products at the high Dp, high ester end.

GLYCEROL

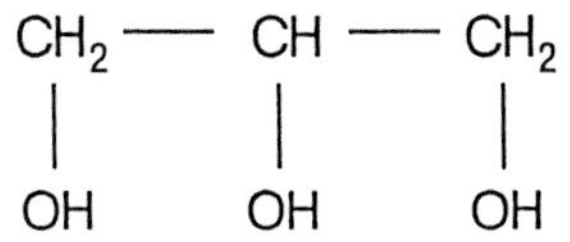

$$\underset{OH}{CH_2} - \underset{OH}{CH} - \underset{OH}{CH_2}$$

LINEAR POLYGLYCEROL

$$HO - \left[CH_2 - \underset{OH}{CH} - CH_2 - O \right]_N - H$$

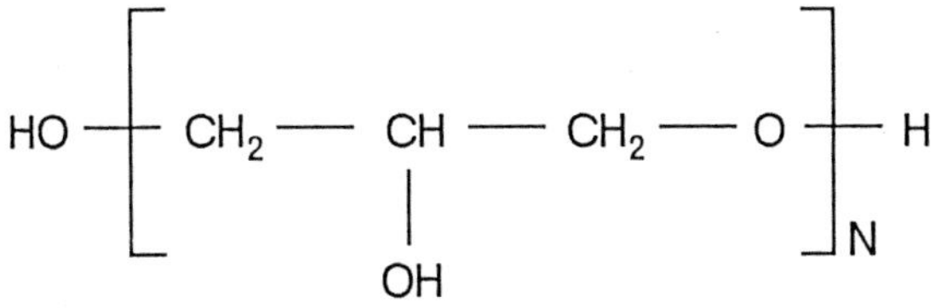

POLYGLYCEROL ESTERS (PGE)

	HYDROXYL GROUPS	M.W.	OHV.
GLYCEROL	3	92	1,826
LINEAR POLYGLYCEROL			
n = 2	4	166	1,349
3	5	240	1,167
4	6	314	1,070
5	7	388	1,010
6	8	462	970

Figure 6

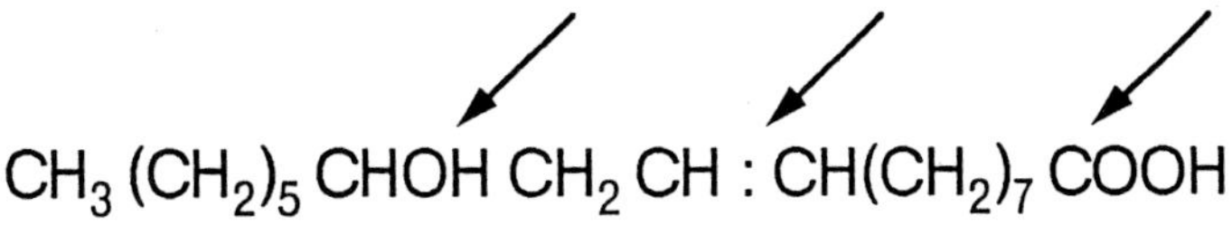

$$CH_3\,(CH_2)_5\,CHOH\;CH_2\,CH : CH(CH_2)_7\,COOH$$

CIS 12 HYDROXY OCTADEC - 9 ENOIC ACID

Figure 7 **RICINOLEIC ACID** (CASTOR OIL FATTY ACID)

Polyglycerol esters have been categorized in terms of these two parameters, e.g.

- 4 - 1 - S tetraglycerol mono stearate
- 6 - 2 - P hexaglycerol di palmitate
- 2 - 4 - O diglycerol tetra oleate

One specific polyglycerol ester worthy of special note, both because of its chemistry and commercial application, is polyglyerol polyricinoleate. Chocolate manufacturers use this ester, usually 0.2 - 0.4% synergistically with lecithin to reduce yield value [viscosity] of their formulations. By its use chocolate formulation can be cheaper, same viscosity at lower fat level, and equally important the flow properties are tailored for particular applications.

Unlike other polyglycerol esters, polyglyerol in this instance is esterified with a prepolymer polyricinoleic acid. Ricinoleic acid [castor oil fatty acid] is 12 hydroxy oleic acid. It has both carboxylic acid and hydroxyl functionality, as well as unsaturation on the 9th carbon. [Figure 7]. Inter-esterified ricinoleic acid is condensed with polyglycerol to make the specific ester, but this product has an even greater number of possible moieties. Polyglyerol polyricinoleate, like other commercial polyglyerol esters, is a mixture, the science being to establish most functional moieties and directing synthesis to their production.

The commercial development of polyglycerol esters has probably been retarded by both their cost and relative difficulty of use, also in some instances organoleptic limitations. From our discussion of their synthesis the cost issue would be apparent, also they are produced on a smaller scale. Polyglycerol esters are indicated for use in confectionery, peanut butter and as an anti-splattering additive in margarine. Perhaps their most appealing attribute is high whippability; an ability to entrain substantial air volumes producing uniform and stable cell structure. Application thus is indicated in imitation cream and fillings, ice cream, cake and particularly sponge batters. Polyglycerol esters can be liquid through waxy and solid but, in general, are not readily water dispersable.

They are not particularly suited to dry mix applications, usually being incorporated in a shortening, emulsifier hydrate, or integrated emulsifier [spray dried or spray cooled]. In such instances a secondary, more hydrophilic emulsifer, e.g. SSL, can be used to aid dispersion and promote the PGE functionality.

(vi) Propylene Glycol Esters of Fatty Acids

Only PGMS, propylene glycol mono stearate, can be considered to have commercial success. Products on the market are 40 - 90% monoesters, although the higher products are considered to have greater functionality. PGMS is prepared by direct esterification of propylene glycol with stearic acid and, like glycerol esters, an excess of propylene glycol directs the reaction to a higher monoester product. The very high, 90% monoester products, are manufactured by distillation.

Like polyglycerol esters, whippability is the most important feature of PGMS. They are waxy solids, with a melting point of approximately 40°C, and hence have similar application difficulties to PGE. Means of formulating around this problem were discussed under PGE and can be similarly applied to PGMS.

(vii) Sucrose Esters of Fatty Acids

Sucrose, being a natural polyhydric alcohol, was an obvious potential for the development of food esters. A disaccharide of glucose and fructose with eight available hydroxy groups, considerable variation of functional properties could be envisaged depending on the degree of esterification. In addition sucrose esters are non-toxic, biodegradable and pass through the body without being absorbed (2).

Commercially, sucrose esters have had limited success. Manufacturers with a basic position in other fatty acid esters have not added sucrose esters to their range, while other manufacturers and countries, e.g. Japan, have focused on them specifically. As a generalization, other fatty acid esters are functionally more cost efficient, sucrose esters merely diluting the market.

Initial efforts to synthesize sucrose esters used a solvent-based [Dimethyl formamide] inter-esterification process with sucrose and methyl esters of fatty acids. More recent developments are commercial, non-solvent processes, using sucrose and either methyl or glycerol esters of fatty acids.

A more comprehensive review of this technology is available in Food Emulsifiers, Developments in Food Science 19, 1989, Charalambous & Doxastakis. (2).

(viii) Acetic, Lactic, Citric and Tartaric Esters of Fatty Acid Monoglycerides

Esterification of a monoglyceride with one of these food acids expands the range of permitted food emulsifiers. The chemistry is essentially similar with the terminal hydroxy group in the fatty acid monoglyceride utilizing carboxylic acid groups in the chosen food acid. Lactic acid can be considered an exception since, analogous to fatty acid lactylates [Section (iv)], reaction conditions favouring lactylic acid formation increase functionality of the end product. Again, choice of fatty acid and degree of esterification are variables, but the majority of commercial products are based on saturated high monoglyceride intermediates.

While the availability of these esters offers the food technologist increased flexibility in emulsifier selection and formulation, again we are dealing with the more specialized, lower volume end of the market. These esters have found applications as follows:

- Acetylated Monoglycerides : Moisture and air barrier - confectionery
  Aeration - whipped toppings
- Citric Acid Monoglycerides : Fat modification - anti-splattering
  Dispersion - coffee whiteners
  Aeration - whipped toppings
- Lactylated Monoglycerides : Aeration - whipped toppings
  - cake and sponge batters

APPLICATIONS

In discussing the broad categories of food esters and their chemistry, it has been necessary to refer to their application and commercial relevance. Rather than repeat and merely catalogue these applications, the following attempts to provide a structured view of surfactant functionality and factors governing choice in a particular food system.

In so doing it is important to realize that no surfactant uniquely displays a specific functionality, rather combinations, also that food systems are both complex and variable. The practising food technologist will use "art" as well as science to reach a satisfactory formulation.

Five major functions can be attributed to food ester surfactants:

1. Emulsification.
2. Aeration.
3. Crystal modification.
4. Starch complexing.
5. Dough conditioning.

Most commercial relevant applications utilize these functions, although others - wetting, dispersing, antifoam, defoam, protective coating, lubricant and viscosity modifier - have also been discussed.

1. **Emulsification**

Fatty acid esters have both hydrophilic components, hydroxyl groups and lipophilic fatty acid groups. Their molecular alignment in water and oil phases respectively reduces interfacial tension and gives rise to their emulsifier function. Ice cream has been mentioned as an oil in water emulsion and margarine water in oil. Fatty acid esters exhibit a wide range of HLB values, but the complex nature of foodstuffs renders HLB a somewhat unreliable guide.

2. **Aeration**

Most glycerol and propylene glycol esters are described as alpha tending which, in whipped emulsions, causes an agglomeration of fat globules and stable foam. Products such as whipped toppings and imitation cream are prepared as emulsions, then aerated with the surfactant aiding both volume and foam stability.

3. Crystal modification

Fats, when rapidly cooled, crystallize in the unstable alpha form, then revert to the more stable beta form. This reversion is described as "bloom" in compounded chocolate and can be retarded by crystal modification using sorbitan tristearate. Fatty esters themselves are kept in the more functional alpha crystal form using a secondary surfactant.

4. Starch complexing

In baked goods, particularly bread, staling is due to crystallization or retrogradation of the amylose fraction in starch. Surfactants with the capability to complex amylose have significant value as anti-staling or crumb softening agents. Saturated high monoglycerides and SSL are particularly valuable in this regard.

5. Dough conditioning

During processing dough is strengthened by the development of a protein network. Surfactants, such as DATA esters and SSL, interact with gluten to reinforce this development, thereby aiding gas retention, increasing volume and improving texture.

Commercial surfactant types are reviewed simplistically by function in the following Table :

	Emulsification		**Aeration**	**Starch Complexing**	**Dough Condit-ioning**
Monoglycerides -					
90% saturated	Good	[W/O]	Excellent	Excellent	Poor
90% unsaturated	Excellent	[W/O]	Poor	Poor	Poor
40-60% saturated	Good	[W/O]	Good	Poor	Poor
Sorbitol Esters -	Poor		Poor	-	-
DATA Esters -	Excellent	[O/W]	-	Moderate	Excellent
SSL -	Excellent	[O/W]	Good	Good	Good
PGE -	Excellent	[O/W]	Excellent	Poor	Poor

This Table is presented as a guide, not a definitive statement. It is based on practical experience, but cannot envisage every possible situation and formulation. To use a simple example, bread making, where starch complexing and dough conditions are the relevant issues. SSL, DATA ester is best for volume and 90% monoglyceride for anti-staling; SSL a reasonable compromise. Such choices do not consider consumer preference on other loaf attributes, flour quality, or processing equipment.

It has been inferred that external factors, not just surfactant function, will influence the choice of additive. For the sake of completeness some of these factors are catalogued:

Consumer based criteria	- product attribute [eg texture] - shelf-life [eg anti-staling] - dietary [eg vegetable/animal derived] - labelling - religious
Raw materials	- variation by season and area
Legal	- permitted levels effective versus legal levels
Cost	-
Processing equipment	- type of addition
Surfactant introduction	- direct addition - dry mix system [bread improver] - fat/shortening - emulsifier hydrate - emulsion

Finally, it should be noted that the range of products and the technology are not restricted solely to the food industry. Some of the product types described also find application in cosmetics, pharmaceuticals and food contact polymers, where the same ethical considerations apply. Food contact polymers in particular have considerable overlap at both product and technology level.

REGULATIONS

Many countries have specific legislation regarding food additives. The main regulations governing food addition are:

JECFA

The Joint Expert Committee on Food Additives operates under the auspices of FAO / WHO.

JECFA processes applications for products proposed as food additives on the basis of toxicology and acceptable daily intake [ADI]. National authorities are unlikely to progress regional applications without JECFA endorsement.

and

European Council Directive 74/329/EEC [as amended].

In addition there is the

Food Chemical Codex

These are analytical standards which regulate purity. Manufacturers must warranty FCC conformity for their products to be accepted by the food industry. It is interesting, as an aside, that while chemical parameters are used to control both process and product, they are not a completely reliable guide to functionality.

BIBLIOGRAPHY

(1) R T McIntyre, Polyglyercol Esters, J.Am. Oil Chem. Soc., Volume 56, 1979.

(2) Allen J. St. Angelo and John R. Vercellotti, Phospholipids and Fatty Acid Esters of Alcohols. Food Emulsifiers, Developments in Food Science 19, Editors G. Charalambous and G. Doxastakis 1989.

(3) Food Emulsifiers, Developments in Food Science 19, Editors G. Charalambous and G. Doxastakis 1989.

(4) CSIRO, Proceedings Fourth Food Industry Conference, Adelaide 1990. Chemicals in Food: Wanted and Unwanted.

(5) Dictionary of Food Ingredients, 2nd Edition 1989, Robert S. Igoe. Publisher - Van Nostrand Reinhold.

(6) Food Additives Handbook, Richard J. Lewis Sr., 1989. Publisher - Van Nostrand Reinhold.

Subject Index